Lecture Notes in Artificial Intel

Edited by J. G. Carbonell and J. Siekmann

Subseries of Lecture Notes in Computer Science

Paul Davidsson   Brian Logan
Keiki Takadama (Eds.)

Multi-Agent and Multi-Agent-Based Simulation

Joint Workshop MABS 2004
New York, NY, USA, July 19, 2004
Revised Selected Papers

 Springer

Series Editors

Jaime G. Carbonell, Carnegie Mellon University, Pittsburgh, PA, USA
Jörg Siekmann, University of Saarland, Saarbrücken, Germany

Volume Editors

Paul Davidsson
Blekinge Institute of Technology
Department of Systems and Software Engineering
37225 Ronneby, Sweden
E-mail: Paul.Davidsson@bth.se

Brian Logan
University of Nottingham
School of Computer Science and IT
Jubilee Campus, Wollaton Road, Nottingham NG8 1BB, UK
E-mail: bsl@cs.nott.ac.uk

Keiki Takadama
Tokyo Institute of Technology
Interdisciplinary Graduate School of Science and Engineering
Dept. of Computational Intelligence and Systems Science
4259 Nagatsuta-cho, Midori-ku, Yokohama, 226-8502 Japan
E-mail: keiki@dis.titech.ac.jp

Library of Congress Control Number: 2005921644

CR Subject Classification (1998): I.2.11, I.2, I.6, C.2.4, J.4, H.4

ISSN 0302-9743
ISBN 3-540-25262-2 Springer Berlin Heidelberg New York

Springer is a part of Springer Science+Business Media

springeronline.com

© Springer-Verlag Berlin Heidelberg 2005
Printed in Germany

Typesetting: Camera-ready by author, data conversion by Scientific Publishing Services, Chennai, India
Printed on acid-free paper   SPIN: 11404668   06/3142   5 4 3 2 1 0

Preface

This volume presents revised and extended versions of selected papers presented at the Joint Workshop on Multi-Agent and Multi-Agent-Based Simulation, a workshop federated with the 3rd International Joint Conference on Autonomous Agents and Multiagent Systems (AAMAS 2004), which was held in New York City, USA, July 19–23, 2004. The workshop was in part a continuation of the International Workshop on Multi-Agent-Based Simulation (MABS) series. Revised versions of papers presented at the four previous MABS workshops have been published as volumes 1534, 1979, 2581, and 2927 in the Lecture Notes in Artificial Intelligence series.

The aim of the workshop was to provide a forum for work in both applications of multi-agent-based simulation and the technical challenges of simulating large multi-agent systems (MAS). There has been considerable recent progress in modelling and analyzing multi-agent systems, and in techniques that apply MAS models to complex real-world systems such as social systems and organizations. Simulation is an increasingly important strand that weaves together this work. In high-risk, high-cost situations, simulations provide critical cost/benefit leverage, and make possible explorations that cannot be carried out in situ:

- Multi-agent approaches to simulating complex systems are key tools in interdisciplinary studies of social systems. Agent-based social simulation (ABSS) research simulates and synthesizes social behavior in order to understand real social systems with properties of self-organization, scalability, robustness, and openness.
- In the MAS community, simulation has been applied to a wide range of MAS research and design problems, from models of complex individual agents employing sophisticated internal mechanisms to models of large-scale societies of relatively simple agents which focus more on the interactions between agents.
- For the simulation community, MAS-based approaches provide a new way of organizing and managing large-scale simulations, e.g., Grid-based simulations, and agent simulation presents a challenging new domain requiring the development of new theory and techniques.

The workshop concerned agent simulation construed broadly, from multi-agent approaches to simulating complex systems, to the simulation of part or all of a multi-agent system and the hard technical issues of multi-agent simulation itself. Contemporary directions in both MABS and MAS research present significant challenges to existing simulation tools and methods, such as concepts and tools for modelling complex social systems and environments; scalability (to thousands or millions of large-grain agents); heterogeneity of simulation components and modelled agents; visualization and steering of simulation behavior;

validation of models and results; human-in-the-loop issues; and more. The workshop provided a forum for social scientists, agent researchers and developers, and simulation researchers to assess the current state of the art in the modelling and simulation of social systems and MAS, identify where existing approaches can be successfully applied, learn about new approaches, and explore future research challenges.

We are very grateful to the workshop participants who engaged enthusiastically in the discussions at the workshop, as well as to the authors' engagement in the second round of review and revision of the papers. We would like to thank Franco Zambonelli, the AAMAS 2004 workshop chair, for having selected the workshop among a large number of high-class proposals. We are also grateful to Nick Jennings and Milind Tambe, the AAMAS 2004 general chairs, for having organized such an excellent conference. Particularly, we would like to express our gratitude to Simon Parsons and Elizabeth Sklar, the AAMAS 2004 local organization chairs, for arranging the infrastructure of the workshop.

Finally, we thank Alfred Hofmann and his team at Springer for giving us the opportunity to continue to disseminate the results of the workshop to a broader audience.

Paul Davidsson, Brian Logan, and Keiki Takadama

Organization

Program Committee

Gul Agha (University of Illinois at Urbana-Champaign, USA)
John Anderson (University of Manitoba, Canada)
Robert Axtell (Brookings Institution, USA)
Rafael Bordini (University of Liverpool, UK)
Francois Bousquet (CIRAD/IRRI, Thailand)
Christopher D. Carothers (Rensselaer Polytechnic Institute, USA)
Shu-Heng Chen (National Chengchi University, Taiwan)
Claudio Cioffi-Revilla (George Mason University, USA)
Helder Coelho (University of Lisbon, Portugal)
Paul Cohen (USC Information Sciences Institute, USA)
Rosaria Conte (IP/CNR Rome, Italy)
Nick Collier (PantaRei LLC/Argonne National Lab, USA)
Daniel Corkill (University of Massachusetts, USA)
Nuno David (ISCTE, Lisbon, Portugal)
Bruce Edmonds (Manchester Metropolitan University, UK)
Richard Fujimoto (Georgia Institute of Technology, USA)
Nigel Gilbert (University of Surrey, UK)
Nick Gotts (Macaulay Institute, UK)
David Hales (University of Bologna, Italy)
Matt Hare (University of Zurich, Switzerland)
Rainer Hegselmann (University of Bayreuth, Germany)
Wander Jager (University of Groningen, Netherlands)
Marco Janssen (Indiana University, USA)
Christophe Le Page (CIRAD, France)
Scott Moss (University of Manchester, UK)
Emma Norling (University of Melbourne, Australia)
Michael North (Argonne National Laboratory, USA)
Mario Paolucci (IP/CNR Rome, Italy)
Alexander Pretschner (Technische Universität München, Germany)
Patrick Riley (Carnegie Mellon University, USA)
Juliette Rouchier (GREQAM (CNRS), France)
Keith Sawyer (Washington University in St. Louis, USA)
Matthias Scheutz (University of Notre Dame, USA)
Jaime Sichman (University of Sao Paulo, Brazil)
Liz Sonenberg (University of Melbourne, Australia)
Takao Terano (University of Tsukuba, Japan)
Georgios Theodoropoulos (University of Birmingham, UK)
Klaus Troitzsch (University of Koblenz, Germany)

Carl Tropper (McGill University, Canada)
Stephen Turner (Nanyang Technological University, Singapore)
Lin Uhrmacher (University of Rostock, Germany)
Harko Verhagen (Stockholm University, Sweden)
Manuela M. Veloso (Carnegie Mellon University, USA)
Regis Vincent (SRI International, USA)
Philip A. Wilsey (University of Cincinnati, USA)

Organizing Committee

Paul Davidsson (Blekinge Institute of Technology, Sweden)
Les Gasser (University of Illinois at Urbana-Champaign, USA)
Brian Logan (University of Nottingham, UK)
Keiki Takadama (Tokyo Institute of Technology, Japan)

Table of Contents

Simulation of Multi-agent Systems

Technique and Technology

Methodology and Modelling

Social Dynamics

Applications

Smooth Scaling Ahead: Progressive MAS Simulation from Single PCs to Grids*

Les Gasser, Kelvin Kakugawa, Brant Chee, and Marc Esteva

Graduate School of Library and Information Science,
University of Illinois at Urbana-Champaign
{gasser, kakugawa, chee, esteva}@uiuc.edu

Abstract. The emerging "Computational Grid" infrastructure poses many new opportunities for the developing science of large scale multi-agent simulation. The ability to migrate agent experiments seamlessly from simple, local single-processor development tools to large-scale distributed simulation environments provides valuable new models for experimentation and software engineering: first develop local, flexible prototypes, then as they become more stable progressively deploy and experiment with them at larger scales. Currently this kind of progressive scalability is hard for both practical and theoretical reasons: Practically, most agent platforms are designed for just one environment of operation. Smooth scalability is more than a matter of increasing agent numbers. Smooth scaling requires clear integration and consistent alignment between a variety of MAS system and simulation architectures and differing underlying infrastructures. This paper reports on recent progress with our experimental platform MACE3J, which now simulates MAS models seamlessly across a variety of scales and architecture types, from single PCs, to Single System Image (SSI) multicomputers, to heterogeneous distributed Grid environments.

1 Introduction

The emerging "Computational Grid" infrastructure [1, 2] poses many new opportunities for the developing science of large scale multi-agent systems. The ability to migrate agent simulations and software seamlessly from simple, local single-processor development tools to large-scale distributed experimental environments will provide valuable new software engineering models for agent-based systems. These new models will couple incremental development of agent simulations with controlled testing and execution. The issue of controlling test and execution is critical because of the complexity of agents as "software engineering units" [3] and because of the need for robustness as experiments are scaled up: the control gained through deterministic simulation is a key engineering tech-

---

* Work supported by NSF Grant 0208937 and by Fulbright/MECD postdoctoral scholarship FU2003-0569.

nique for validating large-scale distributed multi-agent systems and simulations whose behavior can't be easily captured analytically for reasons of complexity.

Scalability is often approached as a problem of size—number of agents, for example. Size is a dominant dimension for some types of large distributed systems. For instance, typical program structures used in large-scale distributed scientific computing—a principal target of the Grid initiatives—employ homogeneous, non-interacting components, such as the typical, regularly-structured "data-parallel" codes found in many data analysis packages, and the assumption that software decision logic is stable while data changes from run to run. In these cases, more processing resources translates directly into the ability to process more data and solve "larger" problems.

For large systems with dynamic, heterogeneous, interacting components such as MAS simulations, scalability raises several other issues. First, scalability requires having appropriate programming models and tools for building and/or integrating many heterogeneous agents. These tools must manage distributed execution resources and timelines, enforce full encapsulation of agents, and offer tight control over message-based multi-agent interactions. The most useful and general programming tools meeting these requirements are based on distributed object models such as the actor model [4]. However, in their pure form distributed object programming models introduce subtle constraints on agent and system construction and operation, which create some difficulties for developing scalable, controllable heterogeneous agent simulations. We detail many of these below.

Second, making simulations smoothly scalable requires aligning simulation models to a variety of programming infrastructures and to a different variety of execution system architectures so that it is possible to conveniently deploy, manage, and control large collections of heterogeneous simulated agents across the different resource pools. For example, simulations designed and run on individual PCs will not have truly concurrent execution, since any single processor necessarily serializes behaviors (see below). In contrast, the so-called Single System Image (SSI) cluster technique binds together a collection of resources to create an abstraction of a distributed system as a unified, single-point-of-access pool of concurrent execution resources or threads. At first, this seems like an almost ideal programming abstraction for concurrent agents, because it hides details of resource allocation and agent deployment. However, there are critical tensions between the monolithic sequential programming approach offered by the single PC environment, and the actual affordances of other distributed platforms across which multi-agent simulations must be developed and deployed to achieve large scales, such as the monolithic concurrent programming approach offered by SSI, and the heterogeneous, distributed, concurrent environment of actual deployed Grids. Grids in existence (e.g., Teragrid [5]) are large and diverse collections of execution resources, with heterogeneous, not regular architecture. Care must be taken to align models based on PC and SSI environments to heterogeneous and fully-distributed environments.

Finally, issues such as infrastructure reliability, functional completeness, and the state of documentation for some kinds of environments including SSI and

Grids can be problematic—Grid services and technologies are a case in point, and this state-of-the-art must be accommodated for realistic agent systems.

In the light of these issues, this paper reports on current developments in our MACE3J experimental platform [6]. The principal innovation reported is the ability to abstract away many of these underlying issues of heterogeneity. This abstraction is accomplished with tools that support seamless deployment of large-grain simulation-based multi-agent systems progressively across a variety of system architectures, ranging from single PCs through SSI clusters to fully distributed Grid environments (in our case based on the Globus Toolkit and the Open Grid Services Initiative) [7, 8]. Below we report on the general structure of the MACE3J system that accomplishes this goal. We next introduce several key dimensions across which multi-agent system simulations need to adapt, and we use these as the basis of a taxonomy of progressively scalable architectures that support MACE3J. We illustrate our progress with reference to several deployed experiments that run across all architectures, and show how making these experiments led to insights about the necessary abstractions.

2   MACE Overview and Design Philosophy

MACE3J is a scientifically oriented multi-agent testbed whose design philosophy is driven by three objectives [6]. We state these here because they are strongly impacted by the need for smooth scalability across architectures.

1. Repeatability and control: MACE3J should support control and randomized repeatability in simulations, which is useful for both development and experimentation.

2. Transitionable models: Agents should be built of components that can be transistioned from simulated implementations or environments to real ones.

3. Generation of knowledge about behavior and structure: MACE3J should support instrumentation that gathers and analyzes data generated by agents and system behaviors.

2.1   Understanding Simulation

A significant focus of MACE3J is simulation support. We view simulation support as the provision of four interlocking types of facilities. MACE3J provides all four of these.

1. *Modeling facilities* capture characteristics of modeled systems such as agents or environments in codes that integrate easily with the activation, coordination, and data-gathering services below. MACE3J generalizes the concept of "agents" to ActiveObjects, which are defined in MACE3J with a set of interfaces. The ActiveObjects concept captures core functionality that allows for implementation of many different types of "agents", so we use the term to denote the foundations for a range of typical agent types. MACE3J modeling facilities include reusable components for constructing

ActiveObjects, environments, and experiments, coupled with the ability to flexibly import these components and models from other projects.

2. ***Coordination facilities*** provide coherence and synchronization for the distributed objects that make up the MAS model. This includes a selectable combination of deterministic (simulation-driven), user-driven, environment-driven, and/or probabilistic control of simulation events, which allow simulations to be re-run exactly, while supporting probabilistic control of behavioral and timing aspects of simulations such as message delay and system failure (e.g., failure of message delivery or of execution). In MACE3J the fundamental coordination object is called `ActivationGroup`. `ActivationGroup` holds a timeline and a set of coordination routines that control the overall execution profile of a simulation.

3. ***Activation services*** that provide enactment (computing) resources for the distributed objects that make up the model. (Of course, these activation services assume an underlying infrastructure such as a single processor PC and SSI cluster, or a Grid environment.) This includes flexible control and steering of simulations through active user involvement in changing simulation parameters at run-time (blurring the distinction between simulation and enactment and facilitating agent transitions to application).

4. ***Flexible data gathering, management, analysis, and presentation*** is done through user-defined and system-defined probes and data streams.

As a simulation development environment, a key objective of MACE3J has been seamless transitioning of models across execution environments. This kind of model retargeting is valuable as an implementation technology and as a software engineering approach: start simple, validate, and expand to more sophisticated environments while exploiting new capabilities. The single processor (possibly threaded) PC platform is stable, well understood, and controllable, but limited in resources. Agent systems can be developed and prototyped rapidly on the single-processor PC platform because it is highly controllable and accepts heterogeneous, changing codes. In contrast, the Grid is less well understood, less flexible, and works best with more homogeneous and stable codes because of the overhead of distribution, startup, and coordination and because of the underlying heterogeneity of the Grid resources. Thus, we need a progression of different development environments and the ability to link them together in a rational, exploratory development process.

The aims of the tools that support such a progressive development approach are these:

1. Provide a simple, direct system model and API to enable maximum flexibility in Agent styles and granularity.
2. Minimize work for users by providing facilities for distributing, deploying, and controlling agent models.
3. Make coordination lightweight, by abstracting the simulation coordination to simple message patterns implemented in infrastructure.

4. Exploit features of existing platforms such as Grid toolkit services to provide agent simulation layer services, to the extent possible. Current examples include deployment services, directory services and communication services.

3   Managing Uncertainty in Scalable MAS Simulations

Here we introduce and develop two main sources of uncertainty for managing design and development of large-scale MAS simulations, with greatest relevance in situations where scale and complexity are related: concurrency and distribution. Concurrency and distribution are inherent properties of MAS, and they introduce several kinds of uncertainty into MAS behaviors.

Concurrency introduces event-ordering uncertainty because concurrently running agents execute at arbitrary rates relative to each other. For a MAS with interactions, increasing scale can increase uncertainty in the ordering of important (interactive) events.

Distribution across space and/or time[1] introduces two types of uncertainty. *Decision uncertainty* occurs when information about the states of remote entities (other agents or environments) that could influence local decisions are inaccessible because those states are distributed. *Semantic uncertainty* occurs when distribution causes agents to translate communicated references or objects into local interpretations that may vary by local context (e.g., [10]). These kinds of uncertainty are fundamental to MAS. Design tools and processes that help control and incrementally modulate distribution- and concurrency-induced uncertainty ease the complexity of engineering and simulating multi-agent systems.

One aim of deployed MAS is to be able to operate in the presence of these types of uncertainty. However, verification of this is hard to do. Our approach is to manage these types of uncertainty by building into middleware support for strategies of progressive, incremental relaxation of control over uncertainty, to gain confidence and experience.

Incremental Management of Event-Ordering Uncertainty: Concurrent execution of agent programs introduces uncertainty about the ordering of interactive agent events such as communications. Event ordering can significantly impact computation results in general, so this uncertainty can have large effects on system reliability, traceability, verifiability, and understandability. Thus, one approach to managing design complexity and improving confidence in MAS behavior is to first eliminate uncertainty in the ordering of events by making events completely repeatable and deterministic. This strict control can then be progressively released to explore system behavior and build confidence under increasing levels of event-ordering uncertainty. In this way, as a MAS experiment is developed, it can be moved to progressively more complex execution environments with progressively greater degrees of freedom in event ordering due to concurrent execution. MACE3J has two ways for exploiting this progressive approach

---

[1] Other dimensions of distribution beyond space and time are also introduced in [9].

to temporal control. First, time- and event-coordinating middleware combines explicit event control with the ability to change architectures. Second, the architectural abstractions of MACE3J allow designers to shift models to progressively more distributed underlying computational architectures that can supply progressively more concurrent resources (such as more processors), and that exhibit behavior closer and closer to the uncertainty of 'real' environments.

Incremental Management of Distribution Uncertainty: Under true distribution, agents may not be able to access information about the internal states of other agents or of their environments. Hewitt called this the problem of "arms-length relationships" [11], and Lesser and colleagues represented it explicitly using partitioned global system models in the TAEMS modeling and simulation approach [12]. A TAEMS model holds an omniscient global view of a problem space and its constraints, while individual agents hold only partial local views. By contrasting the content and accuracy of partial local views with the omniscient global view, an experimenter can measure precisely where her agent control and information sharing strategies have succeeded or failed. Some simulation infrastructures—namely, shared-memory ones—make it far easier than others to model global shared knowledge and partial access to it. However real agent systems, as well as simulations whose aim is to explore actual runtime conditions of distributed agents, cannot rely on shared infrastructure variables. Specifically, multi-agent implementations that rely at all on pointers and/or shared data/variables rather than pure messages for agent interactions will not transfer to truly distributed cases. In this sense, a distributed infrastructure acts as a validation tool for the distributability of agent architectures and agent interactions, keeping them "honest." Thus we can use MACE3J's ability to smoothly scale across multiple infrastructures as a tool for exploring and validating a system's ability to manage distribution-caused uncertainty.

4   Progressive Scaling

In MACE3J we combining support for these two approaches in one middleware layer, allowing designers and experimenters to move a MAS across a following spectrum of environments and control regimes that vary in their degree of actual or apparent concurrency and their degree of distribution. There are six cases:

a) Deterministic single processor, single threaded, shared-memory testbeds which strictly control all temporal progress and inter-agent interactions for completely repeatable performance. In this case, MAS application level coordination mechanisms can be explored, tuned, and verified deterministically against global states, at a cost of realism and the challenge of real variance.

b) Randomized deterministic single processor, single threaded, shared-memory testbeds which, again, control all temporal progress and inter-agent interactions for completely repeatable performance. By randomizing schedules and interaction order, some useful aspects of true concurrency can be achieved statistically. MAS application level coordination mechanisms can again be explored, tuned, and verified deterministically against global states, at a cost of realism and the challenge of real variance.

c) Synchronized multiprocessor, multi-threaded, shared-memory testbeds, in which time and interaction are controlled explicitly and flexibly. In this case, some aspects of time and concurrency (e.g. specific event and interaction types) can be left uncontrolled in MACE3J. Processing concurrency can be exploited to speed up development and testing as needed. The shared-memory aspect still allows for arbitrarily complete control and measurement of any interaction or temporal step that is desired.

d) Unsynchronized multiprocessor, multi-threaded, shared-memory testbeds, in which time and interaction are controlled explicitly and flexibly, and in which MAS application level coordination mechanisms can be explored and verified. In this case, all aspects of time and concurrency must be left uncontrolled. Processing concurrency can be exploited to speed up development and testing as needed. The shared-memory aspect still allows for arbitrarily complete control and measurement of any interaction or temporal step that is desired.

e) Synchronized multi-threaded, distributed processor, *distributed* memory testbeds (e.g., Grid) in which time and interaction are controlled explicitly and flexibly. In this case also, some aspects of time and concurrency can be left uncontrolled. However, the distributed-memory, message-passing aspect adds the requirement of "purer" distributed object techniques (e.g., eliminating shared variables, distilling communications contents to serializable media such as strings). The appropriate middleware infrastructure still allows for arbitrarily complete and repeatable control of any interaction or temporal step, so as to verify behavior and results while approaching realistic execution (temporal, representational, and resource) environments.

f) Unsynchronized multi-threaded, multiprocessor, distributed memory testbeds (Grid) in which time and interactions are controlled only through MAS application level coordination mechanisms. (This is the desired end-state for MAS.)

5   Experiences and Discussion

We have used MACE3J in each of the settings mentioned above for a variety of simulation studies, controlling and progressively releasing various aspects of execution and interaction as described above. Details of some of the sample problem scenarios used can be found in [6]. Currently our most sophisticated agent simulation is the TaskModel experiment. TaskModel takes an arbitrary graph of interdependent tasks (a work ow) and maps them to an arbitrary set of agents for execution. Simulation drives an arbitrarily-timed set of problem – instances through the task network. Agents use a base of local organizational knowledge to reason about task interdependencies, and dynamically reallocates tasks to agents following specifiable task m igration and task allocation policies. Varying any of these dimensions modulates the complexity, difficulty, and scale of the experiment.

Below we discuss several of the key problems that have emerged in using MACE3J across these scenarios and environments, and how we handled and learned from them. For reasons of length, here we focus only on the Grid infrastructure case as it is the most sophisticated of the environments we've explored.

5.1 Grid Rationale and Experiments

Grid services are a new technology for Internet-scale distributed computing [1]. While Grid services are based on the concept of W eb services, Web services are stateless and non-transient, and this makes them ill-suited as hosts for distributed object simulation components such as agents. Grid services, do, however, have several important features that make them particularly suited for distributed computing, including service factories and lifecycle management, which allow distributed objects (agents) to be instantiated and have a controllable level of persistence. Other key advantages of Grid services are their platform- and language-independence, which have always been important considerations in the MACE3J system.

We use Grid services as a distributed computing technology to expose and connect MACE3J `ActivationGroup` and `Agent` objects through the Internet. There are many advantages to this approach. In particular, the toolkit we are using (Globus' GT3 [7]) employs the standardized Grid service specification that covers a wide array of features, including security, indexing, and management of services. In addition, GT3 allows rapid construction of services without the complexity of handling WSDL service descriptions. GT3 encompasses the programming and hosting environment for Grid services, which allows services to be built and deployed in a relatively seamless fashion.

A high-level overview of the mechanics of running an agent experiment can be broken down into three phases. The first phase involves setting up the Grid hosting environment, by deploying the MACE3J services to the Grid installation. The next phase is running the hosting environment and exposing the `ActivationGroup` and Agent services on the Internet. The final phase is to initialize and run a simulation experiment.

5.2 Implementing Agents as "Pure" Distributed Objects

Perhaps the most significant problems we encountered in moving across single PC environments, shared-memory SSI clusters, and distributed Grid environments were failures arising from incomplete or "impure" distributed object models. These led to significant reimplementations in both simulation and middleware components. Four areas of concern arose in our experiments: object identity, inter-object communication, replication, and enforcement of encapsulation.

Identity for Distributed Grid-Based Objects: In the Grid, local objects contain proxy objects that "stand in" for remote partners. Local proxies translate local method calls to remote communications and translate results back. This one-step removal from direct interaction yields a disjointed and asynchronous view of remote agents in the system, fostering difficulties in low-level areas like creation and destruction of agents.

Inter-agent Communication: Inter-agent communication problems arose because of the structure of Messages in original versions of MACE3J. Originally a M essage object encapsulated both pertinent metadata (sender, receiver, time) and message content objects. Using message content objects makes message management much simpler in the local-memory (PC or SSI) infrastructure, and it has the advantage of being able to transfer arbitrary objects—e.g., documents or programs—directly in messages. However, local objects are not communicable in Grid messages, for two reasons. First, message content objects may have arbitrary and context-dependent boundaries (e.g., hash tables with runtime-loaded objects as contents; method or field definitions inherited at runtime) so a message sender must incorporate a dynamic theory of object "edges", and the context for an object is hard or impossible to package and communicate remotely. This means objects sent to remote contexts have uncertain interpretation (execution) semantics when they arrive[2]. The solution is to reduce context by restricting messages to be strings. Abstractly speaking, strings are also freighted with contextual baggage, but the problem of string interpretation is a user-level issue, not a system-level one. Thus we reworked the communication protocol to be more "pure" and use only Strings as media.

Information Replication: The third problem was information replication. In MACE3J, the central axis of coordination for simulation is an object called the **ActivationGroup**. Initially, the single PC version of MACE3J used a static Configuration object that was referenced directly by every part of the system for global configuration values. However, early experiments soon showed this approach to be incompatible with remotely distributed objects that would not share the same program space for configuration. The distributed nature of the Grid system led us to rework how configuration values were accessed by having agents use messages to query the central `ActivationGroup` for configuration values. This in turn led to two types of messages: those for system coordination and those for inter-agent communication.

Encapsulation: Finally, programming support is needed for enforcing encapsulation of agents across all platforms types. This is perhaps the key issue for smooth transitioning of agent types across execution environments. With current widely-used object-based programming environments it is too easy to overlook pointers and object references that extend beyond the encapsulation boundaries of an agent. The instance of context for objects passed in messages, detailed above, is one such encapsulation failure, but the problem is a general one. In the progressive development model, proper encapsulation can be verified at the single-processor environment level so that transitions to the truly distributed environment of the Grid is smooth and seamless.

The limitations these types of problem place on the development of scalable agent systems is that many cross-platform issues have to be taken into consider-

---

[2] C.f., the fundamental issue of *distributed semantics* mentioned above and treated in [10].

ation early in the experiment development process. The result is that transitions to future environment platforms are more easily facilitated with "purer" encapsulation of agents and agent services throughout the system.

6   Conclusions

In conclusion, we find that smooth scalability for multi-agent simulations is not just a size problem—it is also a problem of system architecture style, managing degrees of distribution, and managing degrees of concurrency. The dream of seamless integration across multiple infrastructure scales and programming models is achievable, though current technologies to achieve it are only just emerging and need much more work to be robust. Leading-edge infrastructure like the Grid are still too hard to use, under-documented, and error-prone, but their advantages are becoming clear. The ability to migrate MAS simulations from simple but resource-poor environments to complex, realistic ones, maintaining progressive control over development parameters and building confidence in behavior, can be a valuable strategy for experimentation at large scales.

References

1. Foster, I., Kesselman, C., eds.: The Grid 2: Blueprint for a New Computing Infrastructure. Morgan Kaufmann, Menlo Park, CA (2003)
2. Berman, F., Fox, G., Hey, T.: Grid Computing: Making the Global Infrastructure a Reality. Wiley (2003)
3. Jennings, N.R.: On agent-based software engineering. Artificial Intelligence **117** (2000) 277–296
4. Agha, G.A.: ACTORS: A Model of Concurrent Computation in Distributed Systems. MIT Press, Cambridge, MA (1986)
5. Teragrid: http://teragrid.org/ (2004)
6. Gasser, L., Kakugawa, K.: MACE3J: Fast flexible distributed simulation of large, large-grain multi-agent systems. In: Proceedings of AAMAS. (2002) 745–752
7. Globus: The globus toolkit 3 programmer's tutorial (2003) Version 0.2.2, http://www.casa-sotomayor.net/gt3-tutorial/index.html.
8. Tuecke, S., Czajkowski, K., Foster, I., Frey, J., Graham, S., Kesselman, C., Maguire, T., Sandholm, T., Vanderbilt, P., Snelling, D.: Open grid services infrastructure (OGSI) version 1.0. Technical report, Global Grid Forum (2003) Global Grid Forum Draft Recommendation.
9. Bond, A.H., Gasser, L.: An analysis of problems and research in DAI. In Bond, A.H., Gasser, L., eds.: Readings in Distributed Artificial Intelligence. Morgan Kaufmann Publishers Inc., Menlo Park, CA (1988) 3–35
10. Gasser, L.: Boundaries, identity and aggregation: Plurality issues in multi-agent systems. In Demazeau, Y., Werner, E., eds.: Decentralized Artificial Intelligence III. Elsevier (1992) 199–212
11. Hewitt, C.: The challenge of open systems. Byte Magazine **10** (1985) 223–242
12. Lesser, V., Decker, K., Wagner, T., Carver, N., Garvey, A., Horling, B., Neiman, D., Podorozhny, R., NagendraPrasad, M., Raja, A., Vincent, R., Xuan, P., Zhang, X.: Evolution of the GPGP/TAEMS domain-independent coordination framework. Autonomous Agents and Multi-Agent Systems **9** (2004) 87–143

Agent Communication in Distributed Simulations

Fang Wang, Stephen John Turner, and Lihua Wang

Parallel & Distributed Computing Centre,
School of Computer Engineering, Nanyang Technological University,
639798 Singapore

Abstract. Multi-Agent Systems (MASs) provide a valuable tool for handling increasing software complexity and supporting rapid and accurate decision making.
Various environments for testing, analyzing and developing MASs have been developed. This paper describes an approach to integrating agents into distributed
simulations. Using the JADE toolkit and the HLA (High Level Architecture), a
general architecture is obtained, where both the high level agent specific services
and the underlying middleware comply with international standards. In this paper,
we show how an MAS may be used to represent entities in a simulation, focusing
on the issue of agent to agent communication, as this is one of the key characteristics of MASs. The causality problem in agent communication is described, and
conditions for ensuring consistency are identified. A prototype system has been
implemented to demonstrate the feasibility of our solution and some experimental
results are presented.

Keywords: Multi-agent systems, distributed simulation, high level architecture,
communication, causality, synchronization.

1 Introduction

Today's software applications are mainly characterized by their component-based structures which are usually heterogeneous and distributed. Agent technology provides a
method for handling increasing software complexity and supporting rapid and accurate
decision making. A number of different approaches have emerged as candidates for the
agent architecture, and at the same time, dozens of environments for modelling, testing
and finally implementing agent-based systems have been developed.

An evaluation comparing many systems for developing software agents can be found
in [1]. Jennings et al. [2] provide a good overview of research and development in the field
of autonomous agents and Multi-Agent Systems (MASs). They summarize some agent
applications covering areas including industry, commerce, entertainment and medicine.
Since the concept of agent has become widely used during the last few years, it has
already become involved with simulations.

Distributed simulation enables participants located in different geographical locations to share a common virtual world, which is called a Distributed Virtual Environment
(DVE). The HLA (High Level Architecture) [3] is an industry (IEEE-1516) standard for
modelling and simulation. It can reduce the cost and development time of simulation
systems and increase their capabilities by facilitating the reusability and interoperability of component simulators. It is increasingly being used in various simulation areas,

P. Davidsson et al. (Eds.): MABS 2004, LNAI 3415, pp. 11–24, 2005.

including education, training, analysis, engineering, entertainment and games. In the HLA, a distributed simulation is called a *federation*, and each individual simulator is referred to as a *federate*, one point of attachment to the RunTime Infrastructure (RTI). A federate can be a computer simulation, it can also be a physical device, a passive data viewer or an interface to a human participant.

There are three main research areas in the cross field of combining MASs with simulations, namely:

- to simulate an agent system in order to learn more about its behavior or to investigate the implications of alternative architectures,
- to use agents as entities in a simulation or virtual environment,
- to utilize agents as a way of controlling simulations and providing services.

Most of the current research in using parallel and distributed simulation techniques in multi-agent and multi-agent-based simulation is addressed to the first area. For example, Uhrmacher and Gugler [4] claim that testing MASs requires distributed parallel simulation techniques that take the dynamic pattern of composition and interaction of MASs into account.

Our work is concerned with the second area, namely to develop autonomous agents for representing entities in distributed simulations. The novelty of our project comes from the way in which autonomous agents are integrated into a distributed simulation, with particular attention being given to the efficient use of HLA services [5]. There are many applications such as battlefield simulations, interactive games, etc. that provide the motivation for this integration.

In this paper, we focus on the issue of agent to agent communication, as this is one of the key characteristics of MASs. The causality problem in agent communication is described, and conditions for ensuring consistency are identified. The rest of the paper is organized as follows: Section 2 describes the benefits of combining MASs and the HLA. Section 3 presents our architecture and classifies the different forms of communication in the system. Causality and message ordering problems are discussed in Section 4. Section 5 illustrates how agent to agent communication can be enabled via the HLA. Section 6 introduces a prototype system, and Section 7 gives some experimental results. Finally, conclusions are given in Section 8.

2 Multi-agent Systems and the HLA

The High Level Architecture (HLA) is an industry (IEEE-1516) standard that is designed to promote the reusability and interoperability of component simulators. The standard comprises three components: the HLA interface specification, the rules, and the object model template (OMT). The interface specification, implemented by the Runtime Infrastructure (RTI), defines how federates interact with the federation, and with one another. The responsibilities of federates and their relationship with the RTI are described by the rules. Using the object model template, each federate defines in its simulation object model (SOM) the objects and interactions that are shared. The Federation Object Model (FOM) defines the overall data to be exchanged between federates during a simulation execution.

The RTI provides facilities for allowing federates to interact with each other, as well as to control and manage the simulation. These facilities are classified into six categories [3]:

- *Federation Management*: allows federates to create and destroy federation executions, and join or resign from an existing federation.
- *Declaration Management*: allows federates to establish their intent to publish object attributes and interactions, and to subscribe to updates and interactions produced by other federates.
- *Object Management*: allows federates to create and delete object instances, and produce and receive individual attribute updates and interactions.
- *Ownership Management*: allows federates to transfer the ownership of object attributes during the federation execution.
- *Time Management*: coordinates the advancement of logical time of the federates, and (if appropriate) its relationship to wallclock time.
- *Data Distribution Management (DDM)*: reduces unnecessary information to be transferred between the federates by filtering out irrelevant data.

In Section 1, three main research areas were identified in the cross field of combining MASs with simulations. In the first area, simulation of MASs, the HLA not only supports the reusability of MASs, but also facilitates their interoperability. Developers may implement the components of a simulation as distinct federates, some of which may be useful in future simulations. Moreover, an application may require different kinds of agent architecture, with different properties. Connecting them together becomes a critical problem. However, using the HLA, it can be achieved by linking the federates together via the RTI to form a single federation.

Andersson and Löf investigated some of the benefits of using the HLA as a conceptual basis for a multi-agent environment [6]. They extended the HLA/RTI with KQML (Knowledge and Query Manipulation Language), an Agent Communication Language (ACL). They developed an air combat scenario to test the environment which can host agents and support communication and information distribution. Logan and Theodoropoulos also discuss the application of distributed techniques to the simulation of multi-agent systems [7]. They present an approach to the distributed simulation of agent systems using the SIM_AGENT toolkit and the HLA interoperability framework [8].

In the second area of agent-based simulation, it is not difficult to understand why distributed simulations need to utilize agent technology. Agents make it possible for distributed simulations to achieve rapid and accurate decision making. With the properties of autonomy, social ability, reactivity and pro-activeness, agents can be used to represent entities in DVEs [9]. Here, the HLA can be used to support component-based development of simulation models. This is even more attractive when using MAS technology, because multi-agent systems are inherently distributed and structured. Moreover, "situated" agents may need to interact with another existing simulator that is not an agent. In this case, the agent federates and the enviroment in which they are situated may be linked easily with the RTI. In other words, combining MAS technology and the HLA provides distributed simulations with intelligence and decreases the complexity of simulation development.

In the third area of utilizing agents as a way of controlling simulations and providing services, agents may be used to control a system [10] or provide some of the services of the RTI, such as the Data Distribution Management services [11].

3   Integration of Agents into Distributed Simulations

Multi-Agent Systems (MASs) have both the traditional advantages of distributed and concurrent problem solving, and the additional advantage of sophisticated patterns of interactions [12]. Some characteristics of MASs that distinguish them from traditional programs are [2]: (1) each agent has only partial information about the environment, i.e., a limited viewpoint; (2) no global system control exists; (3) data is decentralized; (4) computation is asynchronous. To integrate agents into an HLA simulation, different approaches have been made toward building up the overall architecture [6]. For a common and widely adoptable architecture, a fundamental concern is an appropriate middleware constructed between the agents and the RTI.

As shown in Figure 1, we have proposed an architecture to build up a prototype system. JADE [13] is a well-known MAS development kit supporting the FIPA specifications [14], an internationally agreed agent standard. So we selected it to support the agents and their communication using ACL. This standard agent platform also contains some agents including the Agent Management Service (AMS), Directory Facilitator (DF) and Remote Monitoring Agent (RMA) that support general agent specific services for all the other agent entities. A *gateway federate* is developed to allow the agents to be connected via the RTI. The federates are built upon the RTI and they can access the RTI interactively.

In the architecture, every agent is attached to a gateway federate. We can have several agents in a group attached to the same gateway federate or a single agent attached to a separate gateway federate. At the same time, different federates can also be placed in the same machine or different machines. Our agents may be regarded as "situated"

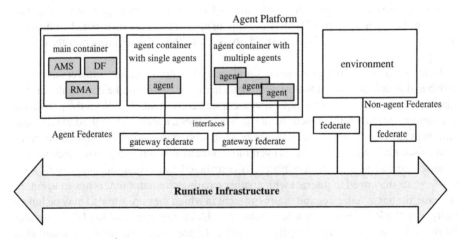

Fig. 1. An architecture for an MAS in an HLA-based simulation

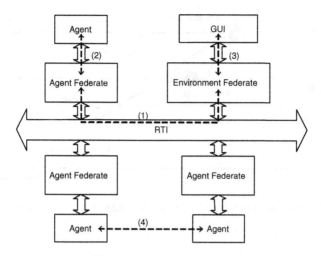

Fig. 2. Classification of communication

or "embedded" agents, that interact with an environment represented by a non-agent federate. In order to enable such interaction, we develop two interfaces to play the role of interaction channels between the gateway federate and the agent, namely, a *sensor* object and an *effector* object. A *sensor* enables an agent to perceive the environment within which it is situated and an *effector* enables it to act upon the environment as a response. Both of them are transferred via the Object-to-Agent (O2A) communication channel provided by the agent toolkit [13]. With these facilities, a complete interaction mechanism for a situated agent comes into being, and a three-phased (*sense-deliberate-react*) agent architecture can be achieved.

Figure 2 provides a conceptual classification of communication in the distributed simulation:

1. *Between different federates*: This includes sending interactions between federates, and producing and receiving individual attribute updates. (e.g. (1) in Figure 2).
2. *Between agent federate and agent*: This includes the *sensor* and *effector* interfaces that are developed for agents to perceive and react. (e.g. (2) in Figure 2).
3. *Between environment federate and GUI*: This provides a way for observers to inter-act with the virtual environment during the simulation execution, for example, by slowing down or speeding up the simulation, or adding or deleting objects in the environment (e.g. (3) in Figure 2).
4. *Between different agents*: This includes sending information in specific formats that can be understood by intelligent agents, such as ACL mesages. (e.g. (4) in Figure 2).

In this paper, we focus on the last form of communication, between different agents in the simulation. However, one of the HLA rules (rule 3) states, "During a federation execution, all exchange of FOM data among federates shall occur via the RTI" [15]. Thus to satisfy the requirements of the HLA and to guarantee the consistency of an

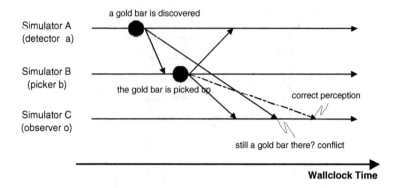

Fig. 3. A scenario of distributed simulation leading to a violation of causality

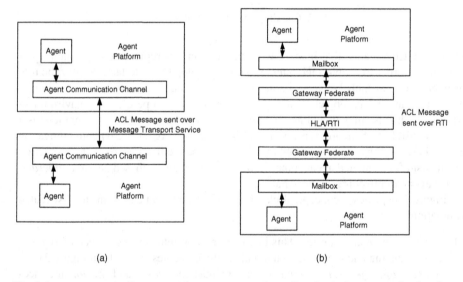

(a)   (b)

Fig. 4. Agent communication route: (a) agents communicate directly via the communication channel provided by the agent platform   (b) agents communicate indirectly via the RTI

application when it is distributed, we must make the communication between agents in different federates indirect via the RTI instead of direct via the communication channel provided by the agent platform (Figure 4). This will be further discussed in Section 5.

As we see, a good agent specific services *middleware* is obtained by combining the agent platform with the gateway federates. Using the JADE toolkit and the HLA, a general architecture is obtained, where both the high level agent specific services and the underlying middleware comply with international standards. Different agent behaviours may be described, making it possible to apply the overall architecture in many other simulations.

4   Causality and Message Ordering

As we know, a basic requirement of a distributed simulation is to ensure that the individual simulators (federates) perceive a common view of the virtual environment. For example, it is important that any pair of participants should perceive a set of messages/events in the same sequence.

Considering a collection of simulators interconnected via a network, it is not difficult to give a scenario leading to a problem of consistency violation without an effective time management system. Suppose that a treasure detector agent a discovers a gold bar, and sends a message to other agents including a picker named b and an observer named o, causing b to pick up this gold bar as a consequence. In the distributed simulation, the messages about the gold bar are as follows: the simulator of a generates a message indicating where the gold bar is. This message is sent to all the other participant simulators including b and o. Upon receiving this message, the gold bar will then be picked up in the simulator of the picker b. Another agent o receives both the messages about the finding from the detector a and the picking up from the picker b. The message about the finding may possibly arrive later than the message about the picking up action, delayed in the network. This causes the observer o to perceive the picking up action first before he knows some gold bar has been discovered. It must be very strange for o that he gets to know a gold bar's location when it has actually been taken away. So is the gold bar still there? This gives rise to a conflict. Thus to keep the causality of the system, messages should arrive in the correct order, that is, the message of discovering a gold bar must arrive before the message of picking up the gold bar. Figure 3 illustrates a scenario about this problem.

This kind of problem is very common in simulations especially when they are distributed. Computer simulations can be classified into discrete models and continuous models, and the most common types of discrete simulations are called *time-stepped* and *event-driven* simulations [3]. They are distinguished by the mechanism used by the simulation to advance simulation time:

- *Time-stepped simulation*: simulation time is subdivided as a sequence of equal-sized time steps, and the simulation advances from one time step to the next.
- *Event-driven simulation*: each event has a time stamp associated with it that indicates the point in simulation time when the event occurs, and the simulation time is advanced to the time of the next event.

Causality can be always kept naturally in the physical world, while in the distributed simulation, it might be affected by many factors such as the latency a message encounters as it is transmitted through the network. Accordingly, a time management mechanism is required to control a distributed simulation to avoid violations of causality. It can also ensure that repeated executions of the simulation with the same initial state and external inputs produce entirely the same result.

The principal types of time management mechanisms are enumerated in [16]. The HLA has a time-stamp order message delivery service. All federates in a federation should explicitly request an advance in their local virtual times. A time advancement request is granted only if no events containing a smaller time-stamp will later arrive at this federate. Furthermore, in a time-stepped simulation, the local virtual time vt advances with a fixed time increment   vt. Generally, in this case, we need some additional conditions from the agent's point of view, such as:

- *An agent should have received all the messages of the last time step before it can process them and advance to another time step.*
- *All the messages that are sent at the same time step vt should also arrive at the same time step* vt+   vt

In this way, as the global time advances, the system's consistency can be well kept between different federates and agents.

5   How Agents Communicate Using HLA

For delivering ACL messages between agents, one possible way is to send them directly to those specified agents listed in the receiver slot, with the facility of the agent toolkit (see Figure 4 (a)). This can be achieved easily no matter whether those agents are in the same physical machine or not. But according to the rule in Section 3 and the discussions in Section 4, this way may cause some consistency violation and is obviously inappropriate.

An alternative way is to send ACL messages indirectly via the RTI utilizing some mechanism (see Figure 4 (b)). In this case, each ACL message can be represented by an interaction sent between federates, and every gateway federate needs a mailbox (MB) as an accessory to take charge of the ACL message transmissions. We develop the mailbox using the JADE agent toolkit and equip it with all the agent specific services. In this sense, we refer to it as a mailbox agent although it is very simple and has only one behavior.

Two constraints need to be made on agent to agent communication, they are:

1. *all outgoing ACL messages of an agent are passed to the local mailbox agent to be processed,* and
2. *a mailbox agent can only send ACL messages directly to local agents that are located in the same federate.*

Those agents in different federates use mailbox agents as communication middlemen and they do not interact with each other directly. These constraints ensure that all exchange of FOM data among federates occurs via the RTI.

5.1   The Mailbox Agent

The mailbox agent has two interfaces connected with its federate. One is *mailboxIN* that contains the received interactions from remote gateway federates, and the other is *mailboxOUT* that contains the interactions to be sent to remote gateway federates. We can see that this is very similar to autonomous agents that have *sensor* and *effector* interfaces with their gateway federates. All the incoming messages collected by a mailbox agent from local agents can be divided into three categories to be processed (Table 1). Here the term *local* indicates that the receivers are in the same gateway federate and the term *remote* indicates the opposite. It is possible that an ACL message is sent to both local and remote agents.

Table 1. Behavior of Mailbox Agent

Receivers	Category	Sent to agent	Sent as interactions
local agents	$isLocal$	yes	no
remote agents	$isRemote$	no	yes
local and remote agents	$isLocal$ & $isRemote$	yes	yes

In summary, when an agent needs to send ACL messages to other agents, a complete transmission procedure consists of following steps:

1. The agent passes the messages to the local mailbox agent (MB).
2. The MB collects all the messages from local agents.
   (a) If the messages are sent to some local agents, the MB sends them accordingly.
   (b) Otherwise,
       i. the MB packs the ACL messages into the *mailboxOUT*, and passes it to the gateway federate.
       ii. The gateway federate reads the *mailboxOUT*, and encodes each ACL message into a byte sequence which is then sent to all the other federates as a time-stamped interaction.
       iii. Another gateway federate receives these interactions, decodes them into ACL messages and packs these messages again into the *mailboxIN* preparing for the subsequent transmission to its MB.
       iv. The MB receives the the *mailboxIN*, and unpacks it into the original ACL messages. The MB then delivers these messages to the local agents that the receiver lists may include.
3. Finally all the destination agents receive the ACL messages.

In this approach, we ensure a federation does not exchange data representing state changes of shared object instances or interactions outside of the RTI service, thus the consistency of the distributed application is not violated.

5.2   Synchronization Between Federates and Agents

In our prototype, the RTI plays the role of the communication system for all the federates, providing the possibility for them to exchange data, and also to synchronize their activities. But the gateway federate and the agent entities still have different threads that proceed concurrently. Thus a problem arises of how to synchronize the threads of the gateway federate, the mailbox agent and the autonomous agents.

For this purpose, a condition variable called latch is introduced here, both between the mailbox agent and the gateway federate, and between the autonomous agent and the gateway federate. Every latch is an object shared by an agent entity and its gateway federate. We let the gateway federate wait on the latch until the agent thread signals it, so that some synchronization points for concurrency are established. Figure 5 shows how a gateway federate synchronizes with its mailbox agent and autonomous agents. The arrows show time dependency, for example, m 3 $\rightarrow$ f3 means $t_{m3} < t_{f3}$, i.e., there exists a time dependency between them. To ensure that all the sent ACL messages have reached the agents before the agents start to process them, we make a restriction, that is, to let the mailbox agent wait until it gets a return receipt for each sent message.

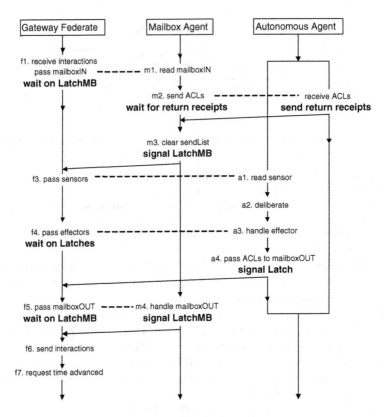

Fig. 5. Synchronization points

6   The Prototype System

Currently a prototype system named MSG (Mine Sweeping Game) has been implemented, in which several soldiers roam to detect and clear all mines out of an area
so as to keep the area away from danger [9]. Figure 6 gives a snapshot of this game.
After picking up a mine, if the soldier's hands become full, he starts to approach the
border to discard the mines. If he is still free to collect more, he can continue to detect
and pick up other mines. A soldier with free hands is able to accept tasks assigned by
other busy soldiers who discover new mines but whose hands are already full at that
time.

The soldiers and the environment are represented by different federates distributed
in a local area network. We can have a group of soldiers executing in the same federate
or make each soldier execute in a separate federate. Different federates may also be in
the same machine or different machines. The federates are connected by the RTI, and
they execute synchronously with each other, controlled by the HLA time management
services. None of the federates can advance its local time until the other federates advance
too. In other words, the present simulation is a synchronous, distributed, time-stepped
system where the simulation advances in equal-sized time steps.

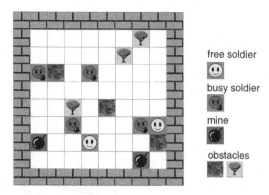

Fig. 6. Six soldiers move in a shared environment avoiding obstacles

The game's main objective is to demonstrate how the architecture enables distributed agents to move in a shared environment performing tasks and to communicate with each other freely. Interaction patterns such as cooperation and collaboration can be adopted in this game too. For cooperation, soldier agents can perform tasks individually and concurrently, each one detecting mines and trying to clear them out. Their common aim is to sweep as many as mines out of the area to reach the final goal that no mine is left, although during this procedure they do not exchange information with each other. This case consists of agents simply working together and no communication exists between agents.

Since communication is one of the key capabilities which distinguishes MASs from other forms of software and provides the underlying power of the paradigm, our soldier agents have been enabled to communicate using identified templates of messages for collaboration. In the current application, soldier agents utilize ACL messages to exchange information. In our MSG policy, we allow a soldier to carry only one mine at a time. That means, after a soldier has discovered and picked up a mine, his task is to reach the border to clear the mine out of the *safe area*. During this period, he is able to detect another mine, however his hands are full at that time, thus he cannot pick it up. What he is able to do then is just inform the other roaming soldiers to come and clear the mine away. In this case, an ACL message carrying the information about the detected mine needs to be sent indicating the position. Obviously, to achieve better performance, the message need only be sent to those soldiers with free hands. Five kinds of ACL templates are utilized by the soldier agents (Table 2).

Table 2. ACL templates applied in the MSG

Template	Implication	State of soldier
informPosition	inform about the position of himself.	unspecified
informPick	inform that he has picked up a mine.	busy
informDiscard	inform about clearing a mine away completed.	free
informMine	inform another soldier about a detected mine's position.	busy
informAck	a return receipt to an *informMine*	free

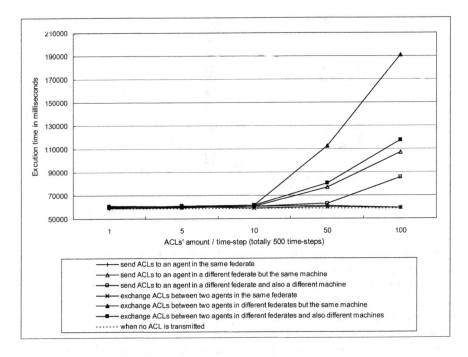

Fig. 7. ACL messages' transmission time

No doubt a certain level of coordination provides a larger viewpoint for every individual agent that has only limited information about the environment, and this always brings a better working efficiency than the situation where no ACL messages and coordination are applied. These issues are discused in further detail in [17].

7   Experimental Results

We have measured the execution times where a number of ACL messages are transmitted between agents in each time step of the simulation. These are tested using three PCs connected via the network. All are of the same type of machine with 2.20GHz Pentium 4 processor and 1.00GB RAM. One of them builds a shared environment for four agents, and the other two support these agents which are located in two different gateway federates. As shown in Figure 7, three situations are investigated for both one-way transmission and two-way transmission.

The execution time does not change much when 1, 5 or 10 ACL messages per time step are transmitted from an agent to another or in both directions at the same time. These results remain stable and are very close to the situation when there is no message transmitted. When the number of messages reaches a large value, such as 50 and 100 messages per time step, the difference among the execution time of the three situations becomes distinct. It should be noted that sending messages to an agent located in the same federate always keeps a stable value of execution time. This is because the messages transmitted *between* federates need to be converted to/from a byte sequence, before the gateway federate can send them as interactions and after receiving them. The encoding and decoding are time-consuming processes that cost computing resources. This also explains why the time of sending messages to an agent in a different federate and also a different machine is less than the time of sending messages to an agent in a different federate but the same machine. The former one does not increase significantly because the encoding and decoding can be carried out in parallel, whereas the latter requires the encoding and decoding to take place on the same machine. For the results, a factor should also be taken into account that messages may encounter network delays when transmitted, however this is not crucial for a local area network.

In all these cases, the gateway federates and the mailbox agents work correctly, according to the conditions for consistency described in Section 4. We can see the transmission of ACL messages is also efficient unless the number of messages sent by an agent is very large, which is not a general case. Actually agents do not need to send messages in each time step of the simulation. Commonly they only communicate if necessary, thus sending ten or even more messages in one time step is unlikely for an agent in practice. However, those cases are still tested as upper bounds. They can also demonstrate that our synchronization mechanisms are quite effective.

8   Conclusions

Although there has been a lot of work on multi-agent and multi-agent based simulation, little has been done on using an MAS to represent entities in a distributed simulation. In this paper, we address the issues of how to integrate agents into an HLA-based distributed simulation with its focus on communication between agents in distinct simulators. We have proposed a general architecture for integrating an MAS in an HLA-based simulation. Both the high level agent platform and the low level infrastructure are standard, thus the overall architecture can be applied in many other agent-based distributed simulations. Adherence to common standards can facilitate the reuse and interoperability of component simulators.

In our approach, a solution is adopted to agent communication using agent gateway federates and mailbox agents to meet the requirements of consistency and causality. A detailed mechanism to provided synchronization between federates and agents is illustrated. A prototype system has been developed to demonstrate the feasibility of this architecture.

So far, we have tested the mechanism and measured the performance. The experimental results show the transmission of ACL messages in the architecture is efficient. In the near future, we plan to investigate more scenarios to advance the proposed agent architecture, and further demonstrate its flexibility and extensibility.

References

1. Eiter, T., Mascardi, V.: Comparing environments for developing software agents. INFSYS Research Report 1843-01-02, Institut und Ludwig Wittgenstein Labor für Informationssysteme, Austria (2001)

2. Jennings, N.R., Sycara, K., Wooldridge, M.: A roadmap of agent research and development. In: Autonomous Agents and Multi-Agent Systems. Volume 1., Kluwer Academic Publishers, Boston (1998) 7–38
3. Fujimoto, R.M., ed.: Parallel and Distributed Simulation Systems. John Wiley & Sons, Inc., Wiley (2000)
4. Uhrmacher, A.M., Gugler, K.: Distributed, parallel simulation of multiple, deliberative agents. In: Proceedings of the 14th Parallel and Distributed Simulation Conferenc (PADS'2000), Bologna (2000) 101–110
5. Wang, L., Turner, S.J., Wang, F.: Interest Management in Agent-Based Distributed Simulations. In: 7th IEEE International Workshop on Distributed Simulation and Real-Time Application (DS-RT 2003), Delft, The Netherlands (2003) 20–27
6. Andersson, J., Löf, S.: HLA as Conceptual Basis for a Multi-Agent Environment. Technical Report 8th-CGF-033, Pitch Kunskapsutveckling AB (1999)
7. Logan, B., Theodoropoulos, G.: The distributed simulation of multi-agent systems. Volume 89. (2001) 174–186
8. Lees, M., Logan, B., Theodoropoulos, G., Oguara, T.: Simulating Agent-Based Systems with HLA: The Case of SIM_AGENT — Part II. In: European Simulation Interoperability Workshop (Euro-SIW), 03E-SIW-076, UK (2003)
9. Wang, F., Turner, S.J., Wang, L.: Integrating Agents into HLA-based Distributed Virtual Environments. In: 4th Workshop on Agent-Based Simulation (ABS2003), Montpellier, France (2003) 9–14
10. Gyurjyan, V., Abbott, D., Heyes, G., Jastrzembski, E., Timmer, C., Wolin, E.: FIPA agent based network distributed control system. In: Computing in High Energy and Nuclear Physics, La Jolla, California (2003)
11. Tan, G., Xu, L.: An Agent-based DDM Filtering Mechanism for HLA. Special Issue on Software Agents and Simulation, Simulation **76** (2001) 329–344
12. Jennings, N.R., Wooldridge, M.J., eds.: Agent Technology: foundations, applications, and markets. Springer. UNICOM, Berlin, Heidelberg and New York (1998)
13. Bellifemine, F., Poggi, A., Rimassa, G.: JADE – A FIPA-compliant agent framework, London (1999) 97–108
14. FIPA: FIPA Agent Management Spacification. Technical Report SC00023J (2002) http://www.fipa.org/.
15. DoD: High-Level Architecture Rules Version 1.3. Technical Report IEEE P1516/D1, U.S. Department of Defense, New York, NY 10017, USA (1998)
16. Fujimoto, R.M.: Time Management in The High Level Architecture. Simulation **71** (1998) 388–400
17. Wang, F., Turner, S.J., Wang, L.: Multi-Agent Interactions in Distributed Virtual Worlds. In: IEEE TENCON 2004. Volume B., Chiang Mai, Thailand (2004) 345–348

Distributed Simulation of MAS

Michael Lees[1], Brian Logan[1], Rob Minson[2], Ton Oguara[2],
and Georgios Theodoropoulos[2]

[1] School of Computer Science and IT,
University of Nottingham, UK
{mhl, bsl}@cs.nott.ac.uk
[2] School of Computer Science,
University of Birmingham, UK
{txo, rzm, gkt}@cs.bham.ac.uk

Abstract. The efficient simulation of multi-agent systems presents particular
challenges which are not addressed by current parallel discrete event simulation
(PDES) models and techniques. While the modelling and simulation of agents,
at least at a coarse grain, is relatively straightforward, it is harder to apply PDES
approaches to the simulation of the agents' environment. In conventional PDES
approaches a system is modelled as a set of logical processes (LPs). Each LP
maintains its own portion of the state of the simulation and interacts with a small
number of other LPs. The interaction between the LPs is assumed to be known in
advance and does not change during the simulation. In contrast, the environment of
a MAS is read and updated by agent and environment LPs in ways which depend
on the evolution of the simulation. As a result, MAS simulations typically have a
large *shared* state which is not associated with any particular agent or environment
LP. In [1] we proposed a new approach to the distributed simulation of MAS in
which the shared state is maintained by a tree of additional logical processes called
Communication Logical Processes (CLP). In this paper we refine this model by
giving precise definitions of a set of operations which allow agent and environment LPs to interact with the shared state and briefly outline how these operations
could be implemented by a CLP.

1 Introduction

Simulation has traditionally played an important role in multi-agent system (MAS)
research and development. It allows a degree of control over experimental conditions
and facilitates the replication of results in a way that is difficult or impossible with a
prototype or fielded system, freeing the agent designer or researcher to focus on key
aspects of a system. As researchers have attempted to simulate larger and more complex
MAS, distributed approaches to simulation have become more attractive [2, 3, 4]. Such
approaches simplify the integration of heterogeneous agents and exploit the natural
parallelism of a MAS, allowing simulation components to be distributed so as to make
best use of the available computational resources.

However the efficient simulation of a multi-agent system presents particular challenges which are not addressed by current parallel discrete event simulation (PDES)

P. Davidsson et al. (Eds.): MABS 2004, LNAI 3415, pp. 25–36, 2005.
© Springer-Verlag Berlin Heidelberg 2005

models and techniques. While the modelling and simulation of agents, at least at a coarse grain, is relatively straightforward, it is harder to apply conventional PDES approaches to the simulation of the agents' environment. Parallel discrete event simulation approaches based on the logical process paradigm assume a fixed decomposition into processes, each of which maintains its own portion of the state of the simulation. The interaction between the processes is assumed to be known in advance and does not change during the simulation. In contrast, simulations of MAS typically have a large *shared* state, the agents' environment, which is only loosely associated with any particular process. The efficient simulation of systems with a large shared state is therefore a key problem in the distributed simulation of MAS.

In [1] we proposed a new approach to the distributed simulation of MAS in which the shared state is maintained by a tree of additional logical processes called Communication Logical Processes (CLP). In this paper we refine this model by giving precise definitions of a set of operations which allow agent and environment logical processes to interact with the shared state and briefly outline how these operations could be implemented by a CLP. In section 2, we present a model of a MAS as a set of logical processes and argue that MAS simulations naturally result in systems with a large shared state. In section 3 we briefly describe our approach to the efficient distribution of the shared state across a tree of CLPs and define a set of operations which allow agent and environment logical processes to access and update the shared state maintained by the CLPs. We then sketch how these operations could be implemented by a CLP, paying particular attention to the problems of efficient sensing, parallel actions and action conflicts. In section 4 we discuss related work and in section 5 we conclude with a brief outline of future work.

2   Modelling a MAS

We adopt a standard parallel discrete event approach with optimistic synchronisation [5, 6]. Decentralised, event-driven distributed simulation is particularly suitable for modelling systems with inherent asynchronous parallelism, such as agent-based systems. This approach seeks to divide the simulation model into a network of concurrent *Logical Processes* (LPs), each maintaining and processing a disjoint portion of the state space of the system. State changes are modelled as timestamped events. Internal events have a causal impact only on the state variables of the LP, whereas external events may also have an impact on the states of other LPs. External events are typically realised as timestamped messages exchanged between the LPs involved.

Agents are *autonomous*. The actions performed by an agent are not solely a function of events in its environment: in the absence of input events, an agent can still produce output events in response to autonomous processes within the agent. As a result, agent simulations have zero lookahead [7]. We therefore adopt an optimistic synchronisation strategy as this theoretically gives the greatest speedup and avoids the problem of lookahead. With optimistic synchronisation, LPs run asynchronously and each has its own local notion of time within the simulation, referred to as its *Local Virtual Time* (LVT). In distributing the simulation across multiple processes a key problem is ensuring that there are no causality violations. An LP is said adhere to the *local causality constraint* (LCC) if it processes all events in nondecreasing time stamp order. If a message arrives in an LP's past (as determined by its LVT) it must rollback its state to the timestamp of the straggler event, and resume processing from that point. It must also cancel any messages it sent with timestamps greater than that of the straggler event, which may in turn initiate rollbacks on other LPs.

We model agents and their environment as Logical Processes. Each agent in the system is modelled as a single *Agent Logical Process* (ALP). Similarly, the properties and behaviour of the objects comprising the agents' environment, e.g., walls, doors, light switches, clocks, etc. and processes not associated with any particular object in the environment, e.g., weather, are modelled as one or more *Environment Logical Processes* (ELP). For example, in a simple Tileworld simulation [8], each Tileworld agent would be simulated by an ALP and the objects in the Tileworld environment (tiles, holes, obstacles etc.) by one or more ELPs.[1] In addition to creating the objects in the environment at simulation startup, the ELP(s) would also be responsible for the creation and deletion of tiles and holes during the simulation. ALPs and ELPs are typically wrappers around existing simulation components. They map to and from the sensor and action interfaces of the agent and environment models to a common representation of the environment expressed in terms of entities and attributes, and also provide support for rollback processing. In what follows we shall use the generic term 'LP' to refer to both ALPs and ELPs, since, unless otherwise noted, their behaviour is very similar.

Each LP has both public data and private data. Private data is data which is not accessible to other LPs in the simulation, e.g., an agent's model of the environment, its current goals, plans etc. or the internal state of a complex object. Public data is data which can, in principle, be accessed or updated by other LPs in the simulation, e.g., the colour, size, shape, position etc. of an agent or object. Public data is held in globally accessible locations or *state variables*, while private data is local to a particular LP. Access to public data and/or the ability to update it may be restricted particular groups of LPs. For example, it may be impossible for any LP to change the size or colour of objects in the environment or for ALPs to update the position of some objects such as obstacles. We model the public data of the LPs in terms of *entities* and *attributes*. We assume each entity in the simulation (agent or object) has a type, and each entity type is associated with a number of attributes. For example, a Tileworld simulation might contain entity types such as *agent*, *tile*, *hole* and *obstacle* and attributes such as *x-position*, *y-position* etc. The shared state of the simulation would therefore consist of a variable number of entities (agents, tiles, holes obstacles etc.) whose properties are defined by the value of their attributes.

In a conventional decentralised event-driven distributed simulation each LP maintains its own portion of the simulation state and LPs interact with each other in a small number of well defined ways. The topology of the simulation is determined by the topology of the simulated system and its decomposition into LPs, and is largely static. In contrast, with multi-agent systems, public data is updated by many LPs and is not logically associated

---

[1] Tileworld is a well established testbed for agents. It consists of an environment consisting of tiles, holes and obstacles, and one or more agents whose goal is to score as many points as possible by pushing tiles to fill in the holes. The environment is dynamic: tiles, holes and obstacles appear and disappear at rates controlled by the simulation developer.

with any of them. Different kinds of agent and environment processes have differing degrees of access to different parts of the environment at different times. In the case of agents, the degree of access is dependent on the range of the agent's sensors (read access) and the actions it can perform (write access). Moreover, in many cases, an agent can effectively change the topology of the simulation, for example, by moving from one part of the environment to another. It is therefore difficult to determine an appropriate topology for a MAS simulation *a priori*, and such simulations typically require a (very) large set of shared variables which could, in principle, be accessed or updated by the ALPs and ELPs.

3   Distributing the Shared State

We therefore propose an approach in which the shared state is loosely associated with a group of special LPs, namely Communication Logical Processes (CLPs), and the distribution of state (i.e., its allocation to CLPs) changes at run-time, in response to the events generated by the ALPs and ELPs during the simulation. Both the allocation of state to CLPs and the synchronisation window are driven by an underlying characteristic of the agent simulation, which we call the sphere of influence [1]. In the Tileworld example above, public data such as the positions of the agents and objects in the environment (tiles, holes and obstacles), the height of the tilestacks, depth of the holes, etc. would be maintained by the CLPs.

ALPs and ELPs interact with the shared state maintained by the CLPs via events, implemented as timestamped messages. The purpose of this interaction is to exchange information regarding the values of those shared state variables which can be accessed or updated by the agent's sensors and actions or by environment processes. Different types of events will typically have different effects on the shared state, and, in general, events of a given type will affect only certain types of state variables (all other things being equal). The 'sphere of influence' of an event is the set of state variables read or updated as a consequence of the event. We can use the spheres of influence of the events generated by each LP to derive an idealised decomposition of the shared state into logical processes (see [1] for details).

3.1   CLPs

The CLPs form a tree with the ALPs and ELPs as the leaves and each CLP maintains a subset of the shared state which is associated with the ALPs/ELPs which are below it in the tree (see Figure 1). CLPs also hold partial information on attributes of entities maintained by other CLPs in the tree, to allow routing of events to the appropriate CLP.

ALPs and ELPs interact with CLPs by exchanging messages. There are 5 message types:

add one or more attributes (and their initial values) at a given timestamp;
remove one or more attributes from a given timestamp;
read the value of one or more attributes at a given timestamp;
write the value of one or more attributes at a given timestamp; and
rollback and resume processing from a given timestamp.

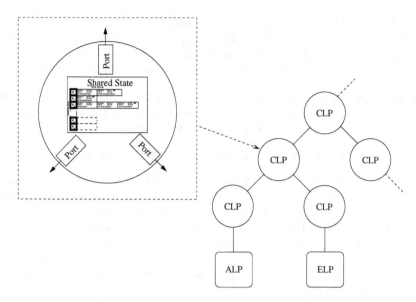

Fig. 1. The tree of CLPs

add, remove, write and *rollback* are non-blocking. A *read* blocks until the requested values are returned. Add, remove, read and write messages originate with an ALP or ELP, while rollbacks are initiated by a CLP. All operations on the shared state occur asynchronously and at the specified simulation time. We assume that the operations are atomic and may be arbitrarily interleaved.[2]

In the remainder of this subsection, we consider each message type in turn and briefly describe their arguments, possible reply messages and any side-effects on the shared state and the state of other LPs. We consider first the case in which the message argument(s) are maintained by the ALP's/ELP's parent CLP. In section 3.2 we describe how messages which can't be handled by the parent CLP are propagated through the tree.

Add Messages. When an ALP or ELP creates a new entity in the simulation its parent CLP adds a new variable to the shared state to hold the value of the attribute. The timestamp indicates the simulation time at which the attribute of the new entity acquired the specified value. Adding the first attribute to an entity instance implicitly creates the entity in the shared state. For simplicity, we assume that entities are only ever created in their entirety, i.e., we cannot create an entity without specifying all values for all its attributes.

Remove Messages. Removing an attribute of an entity in effect deletes the attribute from the specified time forward. Subsequent read and write operations on the attribute with timestamps prior to the specified timestamp proceed as normal. Reads with timestamps

---

[2] The distinction between read and write operations is similar to the query event tagging proposed in [9] and should have similar advantages in reducing both the frequency and depth of rollback and the state saving overhead.

later than the specified timestamp give rise an empty list of values. Attempting to add a new attribute with a timestamp greater than the specified timestamp has no effect (i.e., it is not possible to recreate an attribute after it has been removed from the simulation). As with creation, we assume that entities are only ever deleted in their entirety.

Read Messages. To sense the environment an ALP or ELP it must issue a state query. A *state query* is either a range query (query by attribute value) or an id query (query by attribute id). A *range query* is a list of 4-tuples of the form:

$$< \ entity\text{-}type, \ attribute\text{-}type, \ value\text{-}range, \ timestamp \ >$$

where the *value-range* indicates the attribute values which are of interest (i.e., that match the query). Range queries allow sensing such as 'all tile x-positions within 5 squares'. For example, in a Tileworld simulation, an ALP simulating an agent may issue a range query to discover which tiles are within the sensor range of the agent. Similarly, an ELP responsible for the creation of tiles within a particular region of the Tileworld may issue a range query to check that the cell in which a new tile is to be placed is not currently occupied by an agent (or by a tile created by another ELP and pushed into this region of the Tileworld by an agent).

An id query is a list of 2-tuples of the form:

$$< \ attribute\text{-}id, \ timestamp \ >$$

Id queries allow query by reference, for example, it allows an ALP or ELP to obtain the current value of one of its own public attributes or the current value of an attribute returned by a range query. They are provided as an optimisation for those cases where the attribute in question is guaranteed to persist until after the timestamp of the query.

Reads give rise to a read-response message containing a (possibly empty) set of values (in the case of range queries), or, in the case of an attribute query, a single value. The values returned are those which were valid at the time denoted by the query timestamp. If there is no value with a timestamp equal to that of the query, for example, if the query timestamp lies between the timestamps of two values or is greater than the timestamp of any matching attribute, the read returns the value with the greatest timestamp prior to the timestamp of the query.

Write Messages. When an ALP or ELP updates an attribute of an existing entity, it sends a write message to its parent CLP with a new value and timestamp, indicating the simulation time at which the attribute acquired the specified value. Attribute values are stored in write periods of the appropriate state variable. A *write period* is a logical time interval during which an attribute maintains a particular value. Each write period stores its start and end time, the value of the attribute over that time period, the LP which performed the write and the timestamp of the most recent read by each LP which read the attribute over the time period.[3] New write periods are created when an LP updates the value of an attribute. This splits an existing write period, and triggers a rollback on

---

[3] In practice, not all write periods need to be stored in state variables, e.g., if a write period has a timestamp lower the LVT of any LP it is inaccessible within the simulation and can be fossil collected.

any LPs which read the previous value of the variable at a logical time between the start and end times of the new write period (see below).

In general, there will be a delay between an agent's sensing and action. It is therefore impossible for an LP to know that the state of the environment it sensed before performing an action still holds when the write is performed. We therefore allow write operations to be guarded. A *guard* is a predicate on the shared state in the form of a list of attribute values which must evaluate to true (i.e., the attributes must have the specified values at the timestamp of the write) for the operation to be performed. A guard functions as the precondition for the successful execution of an action in the environment. If the guard evaluates to false, the write is not performed (with the exception that we ignore violations of the precondition due to writes performed by the same agent at the same timestamp). For example, to prevent two (or more) agents pushing the same tile at the same time in Tileworld, we can require that the tile is still where the agent sensed it (e.g., directly in front of the agent) before allowing the agent to update the position of the tile. All writes also have an additional implicit guard, namely that the attribute being updated has not been removed at a timestamp prior to the write.

We distinguish different categories of attributes depending on the types of updates they admit [10]. *Static attributes* are set once, e.g., when an entity is created, and can't be changed during the simulation. Attributes which can be updated at most once at a given timestamp are termed *mutually exclusive attributes*. For example, in Tileworld, we may wish to prohibit two agents picking up a tile at the same time. *Cumulative attributes* can be updated at most n times by different LPs at the same timestamp. For example, in the Tileworld, several agents may be able to drop a tile into a hole at the 'same' time, with each operation decreasing the depth of the hole by one. All updates of static attributes are ignored. If two or more LPs attempt to perform conflicting updates, e.g., attempt to specify different values for a mutually exclusive attribute at the same timestamp or attempt to drop a tile into a hole that has already been filled by other agents at the same timestamp, we apply the update of the LP with the highest rank. The *rank* of an LP determines it's priority when attribute updates conflict. Ranks may reflect some property of the LP which is relevant to the simulation, but in general are simply a way of ensuring repeatability. If both LPs have the same rank then we choose an update arbitrarily (saving the random seed to preserve repeatability). If the attribute has already been updated at this timestamp by an LP with lower rank, this value is over-written and any LPs which read the previous value are rolled back (see below).

More complex environment models can be implemented using combinations of these features. For example, with an appropriate choice of guard on a cumulative attribute, we can allow several agents to push a tile at the same time to give motion which is, e.g., the vector sum of the motion imparted by each agent. Alternatively, an entity's motion can be computed by the ALP or ELP responsible for maintaining the entity in the simulation, with each agent and object updating an input force vector represented as a cumulative attribute.

Rollback Messages. Some sequences of operations by the LPs give rise to further processing of the shared state and the private state of one or more LPs.

An add, remove or write operation with timestamp t which is processed in real time after a read with timestamp t_r, where $t < t_r$, invalidates the read, and triggers a rollback on all LPs which read the previous (interpolated) value of the attribute. A *rollback* indicates that the set of values returned in response to the read was incorrect, and that the LP should rollback its processing to the timestamp of the read and restart.[4] Rolling back an LP undoes all the updates to the LP's private state which have a timestamp $> t_r$ and resets the LP's LVT to t_r. The effect is as if the LP had just returned from the original read (at timestamp t_r), but this time with the 'correct' values of the attributes. (A subsequent add, remove or write with timestamp t', where $t' < t < t_r$ can of course cause further rollbacks on the LP.) Rolling back an LP also cancels any add, remove or write operations on the shared state performed by the LP which have a timestamp $> t_r$. This may in turn invalidate reads by other LPs, requiring them to rollback too.

Note that the presence of rollback obviates the need for coarse-grain atomic operations, i.e., each attribute update can be processed independently of any others and may be arbitrarily interleaved with other operations such as read operations. It is therefore possible for an LP to 'see' an inconsistent version of the shared state or for the guard conditions of a write to evaluate to true for some orderings of operations on the shared state and false for others. When all the updates are finally made, the inconsistency will be detected and any affected LPs rolled back.

3.2   Ports

Each CLP holds only part of the shared state. Read and write operations on shared state variables not maintained by a CLP are forwarded through the tree to the relevant CLP(s).

CLPs communicate with their neighbours in the tree via ports. Each *port* holds information about the ranges of attribute values maintained by CLPs beyond the port in the form of 4-tuples:

$$< \textit{entity-type, attribute-type, value-range timestamp-range} >$$

For example, in a Tileworld simulation, a port tagged with *entity-type* tile, *attribute-type x-position value-range* 10–20 and *timestamp-range* 50–100 would indicate that state variables holding x positions of tiles with values in the range 10 to 20 and timestamps between 50 and 100 are held in CLPs beyond this port. (Where the port leads to an ALP or an ELP, the port information is empty, since all public information in the simulation is held in the CLPs).

In the case of range queries, the query is compared against the range information for each port. If the ranges overlap the CLP forwards the query to the CLP beyond the port. This process proceeds recursively until a CLP with no ports (as opposed to maintained state variables) that match the query is reached. Each CLP waits until it receives replies from all CLPs to which it forwarded the query, appends the value of any state variables it manages that match the query and sends a reply to the originating CLP. When the replies

---

[4] Note that a write with timestamp t_w which arrives in real time after a write with timestamp t'_w, where $t_w < t'_w$ and there are no intervening reads, does not trigger a rollback. In contrast to standard optimistic synchronisation approaches which rollback on every straggler event, or which only avoid rollbacks on straggler reads [9], this optimisation results in a significant reduction in the number of rollbacks [11].

reach the root CLP for this query, the sensing is complete and the values matching the query can be returned to the requesting ALP/ELP.

Initially, the *value-range* for each entity and attribute type at each port is "all values" for all timestamp ranges and all queries are forwarded to all neighbouring CLPs. By analysing the responses to range queries by the neighbouring CLPs, a CLP acquires information about the kinds of attributes (and their ids) that lie beyond each port. This provides a simple form of 'lazy' interest management, and avoids repeated traversal the whole tree when sensing the environment. In addition, each port also holds information about the attribute instances maintained by other CLPs that can be reached via the port. This routing information allows a CLP to forward reads and writes of particular attributes that it does not maintain to the appropriate CLP.

Updating the value of an attribute may involve updating the range information of the ports leading to the CLP which manages the associated state variable. Each CLP keeps a record of all queries it has received together with the port through which the query arrived at the CLP. All add operations are checked against this query history, and, if the new attribute value matches a previously evaluated query, the add is propagated back along the path of the query to update the port information. When the traversal reaches the ALP that initiated the query this triggers a rollback, as the first time the query was evaluated, it returned too few values. Conversely, if an attribute value matches no query in the query record, then no ALP has ever queried this attribute value at this timestamp, and there is no need to propagate the value beyond the current CLP.

3.3   Load Balancing

As well as storing state variables and enabling communication via ports, CLPs also facilitate load balancing. As the number of instances of each event type generated by an ALP or ELP varies, so the partial order over the spheres of influence changes, and the contents of the CLPs must change accordingly to reflect the LPs' current behaviour and keep the communication and computational load balanced. This may be achieved in two ways, namely by swapping pairs of ALPs/ELPs, and by moving subsets of state variables from one CLP to another. In general, it is easier to move state than LPs, and our strategy is to bring the environment close (in a computational sense) to the LPs within whose sphere of influence the corresponding portion of the shared state lies. For example, in a Tileworld simulation, the state associated with entities currently being sensed or manipulated by an agent would ideally be located on the parent CLP of the ALP responsible for simulating the agent. As the agent moves around the Tileworld, the state maintained by the ALP's parent CLP (and its parent CLPs in turn) should change to reflect the agent's changing sphere of influence.

Periodically, the CLPs offer to swap state variables with their neighbours. A CLP will offer to swap a state variable if doing so will reduce the total cost of access. In order to calculate the cost, each query carries with it the 'distance' it travelled through the tree before reaching the CLP. The hop counts for queries arriving through each of the CLP's ports are totalled for each variable maintained by the CLP, and this information is used to determine which port (i.e., neighbouring CLP) to swap with. For example, if the majority of accesses to a state variable arrive through a particular port, a CLP may offer to swap the variable with the CLP which can be reached via the port.

4   Related Work

There is a considerable amount of work in the simulation literature on the efficient distribution of updates, particularly in the context of large scale real-time simulations where it is termed *Interest Management*. Interest Management techniques utilise filtering mechanisms based on *interest expressions* (IEs) to provide the processes in the simulation with only that subset of information which is relevant to them (e.g., based on their location or other application-specific attributes). Special entities in the simulation, referred to as *Interest Managers*, are responsible for filtering generated data and forwarding it to the interested processes based on their IEs [12].

In most existing systems, Interest Management is realised via the use of IP multicast addressing, whereby data is sent to a selected subnet of all potential receivers. A multicast group is defined for each message type, grid cell (spatial location) or region in a multidimensional parameter space in the simulation. Typically, the definition of the multicast groups of receivers is static, based on a priori knowledge of communication patterns between the processes in the simulation [13]. For example, the High Level Architecture (HLA) utilises the *routing space* construct, a multi-dimensional coordinate system whereby simulation federates express their interest in receiving data (subscription regions) or declare their responsibility for publishing data (update regions) [14]. In existing HLA implementations, the routing space is subdivided into a predefined array of fixed size cells and each grid cell is assigned a multicast group which remains fixed throughout the simulation; a process joins those multicast groups whose associated grid cells overlap the process subscription region.

Static, grid-based Interest Management schemes have the disadvantage that they do not adapt to the dynamic changes in the communication patterns between the processes during the simulation and are therefore incapable of balancing the communication and computational load when the communication patterns change, with the result that performance is often poor. Furthermore, in order to filter out all irrelevant data, grid-based filtering requires a small cell size, which in turn implies an increase in the number of multicast groups, a limited resource with high management overhead.

In contrast, our approach is not confined to grids and rectangular regions of multidimensional parameter space and does not rely on the support provided by the TCP/IP protocols. Rather, the shared state is distributed dynamically based on the spheres of influence of the ALPs and ELPs in the simulation. In addition, our approach exploits this decomposition in order to perform load balancing.

5   Conclusion and Further Work

In this paper we have argued that the efficient simulation of the environment of a multi-agent system is a key problem in the distributed simulation of MAS. Building on work in [1], we proposed an approach in which the shared state of a simulation is loosely associated with a group of special logical processes called Communication Logical Processes, and the distribution of state (i.e., its allocation to CLPs) is performed dynamically in response to the events generated by the agent and environment processes during the simulation. We defined a set of operations on the shared state which allow the interaction of agent and environment logical processes and sketched how these operations could be implemented by a CLP. Our approach addresses the problems of efficient sensing, parallel actions and action conflicts, and integrates an efficient approach to state saving which minimises the number of rollbacks with a simple load balancing scheme.

The work reported is still at a preliminary stage. To date, we have implemented the core of the CLPs including the rollback mechanism and calculation of virtual time [15] and load balancing [16] and are currently working on the implementation of interest management. Initial experiments with the rollback mechanism are encouraging, and show a reduction in the number of rollbacks compared to other approaches in the literature which rollback on every straggler event [11].

Acknowledgements

This work is part of the PDES-MAS project[5] and is supported by EPSRC research grant No. GR/R45338/01.

References

1. Logan, B., Theodoropoulos, G.: The distributed simulation of multi-agent systems. Proceedings of the IEEE **89** (2001) 174–186
2. Anderson, J.: A generic distributed simulation system for intelligent agent design and evaluation. In Sarjoughian, H.S., Cellier, F.E., Marefat, M.M., Rozenblit, J.W., eds.: Proceedings of the Tenth Conference on AI, Simulation and Planning, AIS-2000, Society for Computer Simulation International (2000) 36–44
3. Schattenberg, B., Uhrmacher, A.M.: Planning agents in JAMES. Proceedings of the IEEE **89** (2001) 158–173
4. Gasser, L., Kakugawa, K.: MACE3J: Fast flexible distributed simulation of large, large-grain multi-agent systems. In: Proceedings of AAMAS-2002, Bologna (2002)
5. Ferscha, A., Tripathi, S.K.: Parallel and distributed simulation of discrete event systems. Technical Report CS.TR.3336, University of Maryland (1994)
6. Fujimoto, R.: Parallel discrete event simulation. Communications of the ACM **33** (1990) 31–53
7. Uhrmacher, A., Gugler, K.: Distributed, parallel simulation of multiple, deliberative agents. In: Proceedings of Parallel and Distributed Simulation Conference (PADS'2000). (2000) 101–110
8. Pollack, M.E., Ringuette, M.: Introducing the Tileworld: Experimentally evaluating agent architectures. In: National Conference on Artificial Intelligence. (1990) 183–189
9. Sokol, L.M., Briscoe, D.P., Wieland, A.P.: MTW: A strategy for scheduling discrete simulation events for concurrent simulation. In: Proceedings of the SCS Multiconference on Distributed Simulation. SCS Simulation Series, Society for Computer Simulation (1988) 34–42
10. Minson, R., Theodoropoulos, G.: Distributing RePast agent-based simulations with HLA. In: Proceedings of the 2004 European Simulation Interoperability Workshop, Edinburgh, Simulation Interoperability Standards Organisation and Society for Computer Simulation International (2004) (to appear).

---

[5] http://www.cs.bham.ac.uk/research/pdesmas

11. Lees, M., Logan, B., Theodoropoulos, G.: Time windows in multi-agent distributed simulation. In: Proceedings of the 5th EUROSIM Congress on Modelling and Simulation (EuroSim'04). (2004)
12. Morse, K.L.: Interest management in large-scale distributed simulations. Technical Report ICS-TR-96-27 (1996)
13. Morse, K.L.: An Adaptive, Distributed Algorithm for Interest Management. Ph.D. thesis, University of California, Irvine (2000)
14. Defence Modeling and Simulation Office: High Level Architecture RTI Interface Specification, Version 1.3. (1998)
15. Lees, M., Logan, B., Theodoropoulos, G.: Adaptive optimistic synchronisation for multi-agent simulation. In Al-Dabass, D., ed.: Proceedings of the 17th European Simulation Multiconference (ESM 2003), Delft, Society for Modelling and Simulation International and Arbeitsgemeinschaft Simulation, Society for Modelling and Simulation International (2003) 77–82
16. Oguara, T.: Load balancing in distributed simulation of agents. Thesis Report 5, School of Computer Science, University of Birmimgham (2004)

Extending Time Management Support for Multi-agent Systems

Alexander Helleboogh, Tom Holvoet, Danny Weyns, and Yolande Berbers

AgentWise, DistriNet, Department of Computer Science K.U.Leuven, Belgium
{Alexander.Helleboogh, Tom.Holvoet, Danny.Weyns,
Yolande.Berbers}@cs.kuleuven.ac.be

Abstract. Time management is essential when simulating multi-agent
systems (MASs) as it allows consistent and repeatable simulation runs.
So far, time management lacks support to express the timing requirements of a simulation explicitly and at an abstraction level appropriate
for MAS developers. Moreover, integrating time management into a MAS
requires the developer to alter the design of the MAS. In this paper, we
first propose *semantic duration models* to capture timing requirements
that reflect the semantics of MAS activities in an explicit model. Second,
we present a time management infrastructure that starts from a semantic
duration model description to integrate all time management functionality into a MAS transparently, i.e. without requiring the developer to
alter the design of the MAS. We use aspect-oriented programming technology as it allows *separation of concerns*, a crucial software engineering
requirement. As a case, we apply our approach to the Packet-World.

1 Introduction and Problem Statement

Simulation platforms enable multi-agent systems (MASs) to be tested before
they are deployed in the real world. An important requirement for such platforms
is that a MAS can easily be integrated with the simulation infrastructure. The
developers have to be relieved from the low-level technical issues associated with
simulations [1]. This allows the developer to concentrate his or her efforts on the
relevant domain application logic.

An essential technical issue which has to be provided by a simulation platform
is time management [2]. Time management ensures that all temporal characteristics of the problem domain are correctly reproduced in the simulation. Time
management is required in simulation platforms to allow controlled and repeatable simulation runs.

Currently, time management is generally supported by means of time management mechanisms [2, 3, 4] which are built into the simulation platforms. Time
management mechanisms are necessary to enforce all simulation events to be
processed in time-stamp order, irrespective of arbitrary and variable delays in
the execution platform. Examples of time management mechanisms are time-
stepped execution and conservative or optimistic event synchronization mechanisms. When time management mechanisms prevent the execution platform

P. Davidsson et al. (Eds.): MABS 2004, LNAI 3415, pp. 37–48, 2005.

from introducing causality errors, the consistency and repeatability of a simulation can be guaranteed.

Time management is also essential in the context of simulating MASs [5], because timing delays introduced by the underlying execution platform may otherwise affect the simulation results. For example, in [6, 7, 8] it is shown that alterations in the execution platform of the agents can have a severe impact on the simulation behavior of the MAS as a whole, possibly introducing unexpected and unwanted behavior.

MASs allow a system to be modeled at a high level of abstraction. Therefore, it is essential that the support for time management in simulation platforms is raised to an abstraction level appropriate for MAS developers. Currently, time management mechanisms are built into the simulation platforms to hide the technical issues related to maintaining logical time consistency. Nevertheless, a MAS developer is still confronted with a number of unsupported time management issues when simulating a MAS. First, there is a lack of support to express the relation between the activity within a MAS and logical time in an explicit way. Outside a simulation context, the concept of logical time is hardly ever employed: agents are generally not designed as entities maintaining a logical clock and generating time-stamped events. If such systems are simulated, the mapping to logical time has to be tackled by the developer without any support, since time management mechanisms require that the time stamps are already assigned to the events, and only provide support for time stamp ordering. A second problem is the lack of support to integrate all time management functionality into a MAS. Currently, this integration requires the developer (1) to reimplement each agent's actions on the environment to transform them into time stamped events and (2) to direct these events to the simulation platform [9, 10]. Besides the fact that this requires a fair understanding of the simulation platform and its interfaces, it also forces developers to alter the design of the MAS.

This paper describes a way to extend time management support for simulating MASs in order to deal with the problems mentioned above. We give a high-level overview of our approach, based on Fig.1. First, we employ Semantic Duration Models to provide support for the developer to make the timing requirements for the simulation of a MAS explicit. Semantic duration models enable the developer to express the mapping of the activity within a MAS to logical time at a high level of abstraction, allowing the semantic meaning of MAS activities to be taken into account. Second, we describe the Time Management Infrastructure we developed. Our prototype allows time management to be integrated in a MAS transparently, i.e. without requiring the developer to make design changes in the MAS or to have any knowledge from the simulation platform and its interfaces. Our approach employs aspect-oriented programming to achieve separation of concerns. Separation of concerns is important from a software engineering point of view, as this allows all time management functionality needed for simulation purposes to be decoupled from the MAS's functional structure. Based on the description of a semantic duration model, aspect-oriented programming allows time management functionality to be "woven" into a MAS.

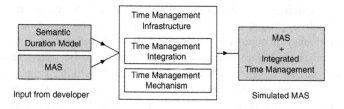

Fig. 1. Overview of the Time Management Infrastructure for simulating MASs

This paper is structured as follows. We first elaborate on semantic duration models in Sect.2 and present a basic formalism based on set theory to describe semantic duration models. Next, the time management infrastructure is described in Sect.3. Section 4 demonstrates our approach using the Packet-World as a case, after which we draw conclusions in Sect.5.

2   Semantic Duration Models

To obtain meaningful simulation results, it is essential that the timing requirements for a simulation reflect the timing characteristics of the MAS's problem domain. We describe how semantic duration models can support a developer to capture all timing requirements of a MAS simulation in an explicit way and at the semantic level of a MAS.

Semantic duration models capture the timing characteristics for simulating a MAS in an explicit way, using the technique of duration modeling at a semantic level. The idea of duration modeling is to maintain a logical clock for each agent and advance that clock for each "primitive" that is executed by the agent. The duration of a "primitive" performed by an agent is the (logical) time period it takes until the effects of that "primitive" are noticeable. The developer has to describe all timing characteristics by means of assigning logical durations to each of the "primitives". Advancing the logical clock in a way that is independent of computer loads and processor speeds, enables repeatable simulation results.

Duration modeling was first described by Anderson and Cohen in [11, 12], where it was applied in the context of the agent's deliberation activity. Anderson distinguishes between low-level and high-level duration models. In low-level models, durations are assigned to individual programming language instructions. However, this results in timing characteristics of a MAS simulation that are described in terms of low-level implementation issues. Because in a problem domain it is the semantics of what the agent is actually doing that determines the timing characteristics, Anderson emphasizes high-level duration models. For example, "evaluating a board position" for a chess playing agent, or "generating an internal plan to reach a particular destination" can be considered as primitives with semantic meaning for duration modeling in a high-level model. However, Anderson's approach is limited to modeling the agent's deliberation activity, and does not take into account other forms of activity within a MAS.

Duration modeling is also addressed in the SPADES system by Riley and Riley [13]. Their approach is not limited to modeling the duration of agent deliberation, but also incorporates the agent's sensing and acting activities. This allows the duration of perception and agent actions to be taken into account. However, in contrast to Anderson's work, the logical thinking time of the agents is now based on the measurement of CPU-time. Moreover, the approach can only be applied to agents whose architecture supports a rigid sense-think-act cycle.

Our notion of semantic duration models combines the best ideas of both approaches described above. First, analogous to the high-level models of Anderson, we consider the "primitives" of duration modeling at the level of activities with a semantic meaning in the behavior of an agent. As a consequence, the duration of each of the activities depends upon the semantic meaning within the context of the simulation only, and is irrespective of the programming language and implementation. Second, analogous to the SPADES system, we extend duration modeling from agent activities employed for deliberation purposes, to activities an agent can perform on the environment. In our semantic duration models, we make a distinction between the agent's internal and external activities. Internal activities are typically related to deliberation and do not cross the agent's boundaries. External activities on the other hand cross the boundaries of an agent and typically include perception of the environment, sending or receiving communication messages and performing actions on the environment. In contrast to the sense-think-act cycle employed in the SPADES system, we impose no order on the agent's internal and external activities.

In our current model, we assume that an agent is the unit of concurrency. As such, each agent can only perform one activity at the same time. However, activities performed by different agents can of course be concurrent.

We describe semantic duration models using a basic form of set theory:

$A = \{a_1, a_2, \ldots, a_n\}$, the set of all agents in the MAS:
$\forall a_i \in A$:
$\quad D_i = \{d_1^i, d_2^i, \ldots, d_{n_i}^i\}$, the set of all internal activities of agent a_i
$\quad E_i = \{e_1^i, e_2^i, \ldots, e_{m_i}^i\}$, the set of external activities that agent a_i
$\quad D_i \cap E_i = \phi$

By combining sets D_i and E_i we obtain:

$\forall a_i \in A$:
$\quad C_i = E_i \cup D_i = \{e_1^i, e_2^i, \ldots, e_{m_i}^i, d_1^i, d_2^i, \ldots, d_{n_i}^i\}$, the set of all activities of a_i
$\quad$ or $C_i = \{c_1^i, c_2^i, \ldots, c_{u_i}^i\}$ with $|C_i| = u_i = m_i + n_i$, the cardinality of C_i

To obtain a semantic duration model for an agent, the duration of all its activities is expressed in terms of logical time. Formally this is equivalent to a function assigning a logical duration to each activity:

$$Duration_i : C_i \times S_i \times W \to \Re$$
$$Duration_i(c_j^i, s_i, w) = r_j^i$$

where S_i is the set of all states of agent a_i, W is the set of all states of the environment, $\Re$ is the set of real numbers and $Duration_i$ is the semantic duration function for agent a_i. $Duration_i$ defines the logical time period it takes until the effects of activity c_j^i performed by agent a_i are noticeable, given that the state of agent a_i is s_i and the state of the world is w. In general, the duration of a particular activity for an agent not only depends on the kind of activity, but also on the state of the agent as well as on the state of the environment.

3 Time Management Transparency

In order to integrate time management into a MAS transparently, the following requirements have to be fulfilled. First, explicit and developer-friendly support for describing semantic duration models must be provided to the developer. The developer should only describe the internal and external activities and their semantic durations (see Sect.2). Based on this, the platform should be able to enforce the time mapping without further intervention from the developer. Second, it must be possible to simulate a MAS without requiring the developer to perform changes in the design of the MAS. However, because time management requires monitoring and controlling the activities of all agents according to user-defined timing characteristics, it requires introducing code in many places across the system. We could refactor all the code and perform the appropriate insertions, but in a large MAS, this would be a time-consuming and error-prone job, which we would like to avoid.

3.1 Aspect-Oriented Programming

Time management is a crosscutting concern, i.e. the time management functionality cross-cuts the MAS's basic functionality. The problem of crosscutting concerns is that they can not be modularized with traditional OO-techniques. This forces the implementation of time management to be scattered throughout the code of the MAS, resulting in "tangled code" that is excessively difficult to develop and maintain. Aspect-oriented programming [14, 15] handles crosscutting concerns by providing aspects for expressing these concerns in a modularized way. An aspect is a modular unit of crosscutting implementation. Aspect-oriented programming does not replace existing programming paradigms and languages, but instead, it can be seen as a co-existing, complementary technique that can improve the utility and expressiveness of existing languages. It enhances the ability to express the separation of concerns which is necessary for well-designed, maintainable software systems.

A language extension to Java which supports aspect-oriented programming, is AspectJ. In AspectJ, defining an aspect is based on two main concepts: pointcuts and advice. A pointcut is a language construct in AspectJ that selects particular join points, based on well-defined criteria. Each join point represents a particular point in the execution flow of a program where the aspect can interfere, e.g. a point in the flow when a particular method is called. As such, pointcuts are a means to express the crosscutting nature of an aspect. Advice on the other hand is a language construct in AspectJ that defines additional code that runs at join points specified by an associated pointcut. An aspect encapsulates a particular crosscutting concern and can contain several pointcut and advice definitions. The process of inserting all crosscutting code of an aspect at the appropriate join points within the original program code, is called aspect weaving. Aspect weaving is performed at compile-time in AspectJ.

3.2   The Prototype

According to the requirements above, we developed a prototype in Java which uses AspectJ to integrate time management as a separate concern. We illustrate its working using Fig.2.

To be able to use time management support, the developer composes a particular Semantic Duration Model Configuration which describes a semantic duration model for each agent within the MAS (see Fig.2). Currently, in our prototype abstraction is made from the state dependency in semantic duration models. As a consequence, $Duration_i$ is simplified to:

$$Duration_i : C_i \to \Re$$
$$Duration_i(c_j^i) = r_j^i$$

This allows $Duration_i$ to be described in terms of a list of (c_j^i, r_j^i)-tuples for each agent a_i, with c_j^i mapping to a Java method that the agent executes to perform a particular activity with semantic meaning, and r_j^i a constant denoting the logical duration of that activity.

After a semantic duration model has been defined for each agent in the MAS, the prototype generates an Aspect and a Time Monitor for each agent. The Time Monitor of agent a_i contains a logical clock for the agent, together with the time mapping as described by $Duration_i$ of that agent (which maps c_j^i to r_j^i). The goal of a Time Monitor is to keep the agent's logical clock up-to-date by advancing it according to the activities the agent decides to perform. When the Time Monitor is notified of the execution of activity c_j^i, it advances its clock by r_j^i. The goal of the Aspect on the other hand is to notify the Time Monitor of all activities the agent executes. Therefore, the Aspect weaves code into all methods that are defined as activities c_j^i of the agent. The goal of the inserted code is to intercept the execution of the agent as soon as it decides to perform an activity and to notify the Time Monitor, such that the agent's logical clock is advanced appropriately. The notification of the Time Monitor by the inserted code is represented graphically by the arrowed lines in Fig.2.

The combination of Aspects and Time Monitors allows the logical clock of all agents to advance according to all executed activities. A MAS Time Synchronizer prevents the occurrence of causality errors. The developer can specify a subset

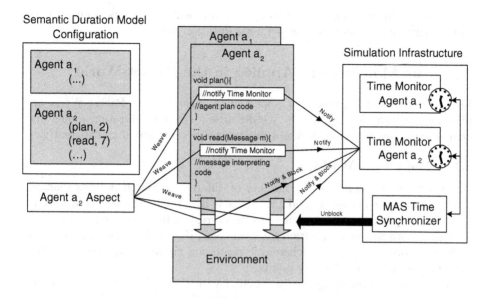

Fig. 2. Time Management Infrastructure for MASs: the gray shaded parts have to be provided by the developer. All white parts are hidden from the developer

of activities that can introduce causality errors and hence have to be controlled by the M A S T im e Synchronizer to ensure that these activities are not executed out of logical clock order. By default, the set of activities for which causality has to be preserved contains all external activities, because these activities cross the agent's boundaries (see Sect.2). In Fig.2, the gray arrows between the agent and the environment represent external activities the agent can perform on the environment.

We explain the approach employed for synchronization by using an example. Suppose that a particular agent decides to perceive its neighboring environment and triggers an external perception activity. The code inserted by the A spect intercepts the execution of the agent right before the chosen activity is actually executed, notifies that agent's T im e M onitor, which advances that agent's logical clock with the appropriate duration and then blocks that agent's execution. Unblocking can only be done by the M A S T im e Synchronizer, which monitors the logical clocks of all agents and employs a conservative time management mechanism [3] to prevent causality errors. The specific way of interception ensures that the logical clocks of the agents are already updated before the corresponding activities are actually executed. This enables the M A S T im e Synchronizer to have prior knowledge of the time stamp of the next activity a particular agent will perform. As a consequence, the synchronization approach applied here does not rely on a lookahead to prevent starvation. In our example, the perception activity of the agent will be unblocked as soon as the M A S T im e Synchronizer can guarantee that all external activities the other agents will perform, have a higher logical time stamp than the perception activity of the former agent.

As such, the former agent perceives the environment in correspondence to the causal order that arises from the semantic duration models.

4   Time Management Applied in the Packet-World

In this section, we illustrate our approach by means of the Packet-World application we have developed [16]. We describe a semantic duration model and demonstrate how time management functionality is integrated transparently.

4.1   The Packet-World

The Packet-World consists of a number of differently colored packets that are scattered over a rectangular grid. Agents that live in this virtual world have to collect those packets and bring them to the correspondingly colored destination. The grid contains one destination for each color. Figure 3 shows an example of a Packet-World with size 10 wherein 5 agents are situated. Squares symbolize packets and circles are delivery points.

In the Packet-World, agents can interact with the environment in a number of ways. We allow agents to perform a number of basic actions. First, an agent can make a step to one of the free neighboring fields around it. Second, if an agent is not carrying any packet, it can pick one up from one of its neighboring fields. Third, an agent can put down the packet it carries on one of the free neighboring fields around it, which could of course be the destination field of that particular packet. It is important to notice that each agent of the Packet-World has only a limited view on the world. This view only covers a small part of the environment around the agent (see Fig.3). Furthermore, agents can interact with other agents too. We allow agents to communicate with other agents by sending messages. In this way, agents can inform each other about the position of packets and destinations. All action and message handling is performed by the environment.

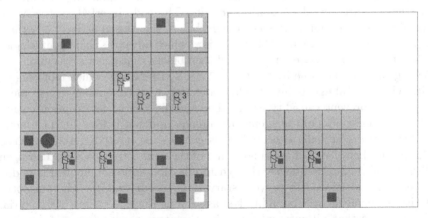

Fig. 3. The Packet-World: global screenshot *(left)* and view range of agent nr.4 *(right)*

4.2   Timing Requirements for the Simulation

In the Packet-World, each agent is an autonomous and pro-active entity which continuously deliberates and invokes actions in the environment. Neither time-stamps, nor events are employed in the agents' design. However, for our simulation, we would like the agents to behave according to specific timing characteristics. Suppose we impose the following timing requirements on the agents: first, picking up or putting down a packet only takes half the time for an agent than performing a step. On the other hand, obtaining perception of the environment or retrieving messages which have arrived, can be done instantaneously. The time it takes for an agent to analyze its perception cannot be neglected. Searching for a destination field based on the input obtained from perception takes as long for an agent as performing a pick up packet action, while finding the nearest packet based on its perception only takes half as long. The time it takes for an agent to select its next action is equal to that of performing a move. Finally, sending a message is twice as costly as performing a step.

4.3   Defining a Semantic Duration Model

We identify all agents' activities in the Packet-World simulation. Using the description above, we can distinguish the following external activities on the environment: an agent can (1) look to perceive its surroundings, (2) move, (3) pick up a packet, (4) put down a packet, (5) send a message, and (6) receive messages that have arrived. Formally (see Sect.2):

$$\forall a_i \in A :$$
$$E_i = \{look, move, pick, put, send, receive\}$$

With respect to the internal activities of the agents, in our simulation a distinction is made between (1) detecting a destination, (2) finding the nearest packet and (3) selecting the next action. Formally:

$$\forall a_i \in A :$$
$$D_i = \{detectdest, findpacket, selectaction\}$$
$$\text{and } C_i = \{look, move, pick, put, send, receive, detectdest, findpacket,$$
$$selectaction\}$$

To define a semantic duration model, we have to assign a duration to each of the activities of an agent, according to the timing requirements of the simulation. We get:

$$\forall a_i \in A :$$
$$Duration_i(move) = Duration_i(selectaction) = 1$$
$$Duration_i(pick) = Duration_i(put) = Duration_i(detectdest) = 0.5$$
$$Duration_i(look) = Duration_i(receive) = 0$$
$$Duration_i(findpacket) = 0.25$$
$$Duration_i(send) = 2$$

Note that the absolute values of the durations are of no importance, only the relative values are significant.

4.4   Integrating Timing Management Code

For each activity described in the semantic duration model of the Packet-World agents, time management code has to be inserted. As an example, we consider the *findpacket* internal activity of an agent (see Fig.4). Based on the semantic duration model described above, an aspect is generated for the *findpacket* activity. The pointcut of the aspect refers to the location of the *findpacket* activity in the agent's code. At this location, the aspect's advice is woven which notifies the agent's time monitor each time the activity is performed.

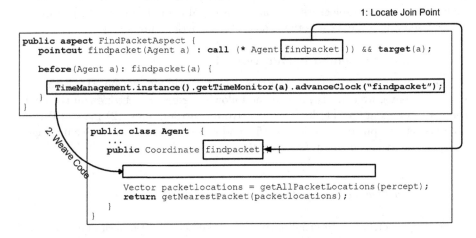

Fig. 4. Aspect weaving for the internal activity *findpacket*

5   Conclusions and Future Work

In this paper, we described a way to extend time management support for simulating MASs. Our contribution consists of two parts.

First, semantic duration models allow the timing requirements of a simulation to be described in an explicit way by means of a user-friendly formalism based on set theory. Semantic duration models employ the technique of duration modeling at a semantic rather than syntactic level and allow timing requirements to be expressed for the internal as well as the external activities of an agent.

Second, we described a time management infrastructure that allows all time management functionality to be integrated transparently in a MAS. The developer describes all timing requirements by means of semantic duration models. To achieve separation of concerns, which is important for well-designed and maintainable software systems, aspect-oriented programming is used. Our prototype allows all time management code necessary for the simulation to be incorporated in the MAS without requiring the developer to change the design of the MAS.

In the paper, we demonstrated our approach in the Packet-World. It was shown that it is possible to control the execution of the simulation according to specific timing requirements and to integrate time management functionality in a transparent way.

Although the approach presented here is promising, a number of issues requires further research and will be addressed in detail in future work.

- With respect to the semantic duration models, we exclusively elaborated upon agent activities, both internal and external. However, activities can also originate from the environment of the MAS, independent of the agents. An example are digital pheromones [17] that propagate and evaporate over time. Pheromones are used for indirect communication in MASs. Our approach requires further investigation with respect to such environmental activities in general.
- In the current model, there is no support to allow overlap of activities, as described in [13]. All activities of an individual agent happen sequentially. An important issue we are currently working on is extending the semantic duration model of an agent such that activities can be specified to be potentially overlapping.
- In our prototype, the current support for semantic duration models is useful but still rather limited, since only constant logical durations can be assigned to activities. Extensions to more complex dependencies are planned in the future.
- Finally, there is no clean duration semantics for hierarchical activities. Suppose agent a_i has two activities: activity c_j^i with a duration of r_j^i and activity c_k^i with a duration of r_k^i, and suppose c_j^i calls c_k^i. If agent a_i then executes activity c_j^i, it is unclear whether agent a_i has to be assigned a logical delay of r_j^i as defined earlier, or $r_j^i + r_k^i$ (which is currently the case in our prototype).

Acknowledgements

This research is partially funded by the KULeuven research project AgCo2 (Agents for Coordination and Control).

References

1. Maria Bruno Marietto, Nuno David, J.S.S.H.C.: Requirements analysis of agent-based simulation platforms: State of the art and new prospects. In: Multi-Agent-Based Simulation, Third International Workshop, MABS 2002. Lecture Notes in Computer Science, Springer-Verlag (2002)
2. Fujimoto, R.: Time management in the high level architecture. Simulation, Special Issue on High Level Architecture **71** (1998) 388–400
3. Chandy, K.M., Misra, J.: Asynchronous distributed simulation via a sequence of parallel computations. Communications of the ACM **24** (1981) 198–205

4. Jefferson, D., Sowizral, H.: Fast concurrent simulation using the time warp mechanism. In: Proceedings of the SCS Multiconference on Distributed simulation. (1985) 63–69
5. Helleboogh, A., Holvoet, T., Weyns, D.: Towards time management adaptability in multi-agent systems. In Kudenko, D., Alonso, E., Kazakov, D., eds.: Proceedings of the AISB 2004 Fourth Symposium on Adaptive Agents and Multi-Agent Systems. (2004) 20–30
6. Axtell, R.: Effects of interaction topology and activation regime in several multi-agent systems. In: MABS. (2000) 33–48
7. Page, S.: On incentives and updating in agent based models. Journal of Computational Economics **10** (1997) 67–87
8. Cornforth, D., Green, D.G., Newth, D., Kirley, M.: Do artificial ants march in step? Ordered asynchronous processes and modularity in biological systems. In: Proceedings of the eighth international conference on Artificial life, MIT Press (2003) 28–32
9. Uhrmacher, A., Kullick, B.: Plug and test software agents in virtual environments. In: Winter Simulation Conference - WSC'2000. (2000)
10. Himmelspach, J., Rhl, M., Uhrmacher, A.: Simulation for testing software agents - an exploration based on JAMES. In: Proc. of the 2003 Winter Simulation Conference, New Orleans, USA. (2003)
11. Anderson, S.D., Cohen, P.R.: Timed Common Lisp: the duration of deliberation. SIGART Bull. **7** (1996) 11–15
12. Anderson, S.D.: Simulation of multiple time-pressured agents. In: Winter Simulation Conference. (1997) 397–404
13. Riley, P., Riley, G.: SPADES — a distributed agent simulation environment with software-in-the-loop execution. In Chick, S., Sánchez, P.J., Ferrin, D., Morrice, D.J., eds.: Winter Simulation Conference Proceedings. Volume 1. (2003) 817–825
14. Kiczales, G., Lamping, J., Menhdhekar, A., Maeda, C., Lopes, C., Loingtier, J.M., Irwin, J.: Aspect-oriented programming. In Akşit, M., Matsuoka, S., eds.: Proceedings European Conference on Object-Oriented Programming. Volume 1241. Springer-Verlag, Berlin, Heidelberg, and New York (1997) 220–242
15. Kiczales, G., Hilsdale, E., Hugunin, J., Kersten, M., Palm, J., Griswold, W.: Getting started with AspectJ. Commun. ACM **44** (2001) 59–65
16. Weyns, D., Holvoet, T.: The Packet-World as a case to study sociality in multi-agent systems. In: Autonomous Agents and Multi-Agent Systems, AAMAS 2002, Bologna, Italy. (2002)
17. Sauter, J.A., Matthews, R., Parunak, H.V.D.: Evolving adaptive pheromone path planning mechanisms. The First International Joint Conference on Autonomous Agents and Multiagent Systems, AAMAS 2002 (2002)

Designing and Implementing MABS in AKIRA

Giovanni Pezzulo and Giangugliclmo Calvi

ISTC-CNR viale Marx, 15 – 00137 Roma Italy
g.pezzulo@istc.cnr.it, calvi@noze.it

Abstract. Here we present AKIRA, a framework for Agent-based cognitive and social simulations. AKIRA is an open-source project, currently developed mainly at ISTC-CNR, that exploits state-of-the-art techniques and tools. It gives to the programmer a number of facilities for building Agents at different levels of complexity (e.g. reactive, deliberative, layered). Here we describe the main architectural features (i.e. Hybridism of the Agents and the Energy Model) and the theoretical assumptions that motivate it. We also present some simulations.

1 Introduction

AKIRA is an open source [16] framework for social and cognitive Agent Based simulations. Many existing platforms for social simulation [17, 18] allow developers to build complex simulations using only very simple Agents. On the contrary, the aim of the Artificial Intelligence Group at ISTC-CNR is to perform social simulations that take advantage of more complex Agents: a theoretical major claim of the Group is that many socio-cognitive phenomena (e.g. involving trust [1] and reputation [2]) can be modeled only with Agents having a certain set of cognitive features (e.g. BDI-like Agents [12]). Another current issue of the Group is to implement and test a range of single-Agent cognitive models (e.g. about beliefs and goals dynamics, expectations, epistemic actions) that were developed in the Institute throughout the years [1, 2, 11].

So, our first requirement for a platform is representing and implementing in the same framework both the cognitive components of single cognitive Agents and their social dynamics, as well as a dynamical environment. AKIRA allows developers to design agents at different levels of complexity: each single cognitive Agent can be implemented either as a *simple-Agent*, e.g. using a single daemon (and a single thread of execution); or as a *complex-Agent*, using many daemons, representing cooperating and concurrent sub-cognitive unities, e.g. goals, beliefs, etc. *Simple* and *complex-Agents* can take a part in a social simulation within the same framework. Moreover, the environment, with its rules and dynamics, can be modeled as an Agent, too, interacting in a transparent way with the other Agents. Some related requirements are: allow both turn-based and real-time (parallel) simulations; furnish utilities for rapid prototyping of different agent based architectures (e.g. reactive, deliberative, layered, modular) models and mechanisms (e.g. BDI, Behavior Networks). At the same time, AKIRA has some built-in properties: Agents can form Coalitions; there is integration of top-down control and emerging behavior; each active Agent introduces a pressure over the whole computation; Agents compete for a limited pool of resources.

P. Davidsson et al. (Eds.): MABS 2004, LNAI 3415, pp. 49–64, 2005.

2 Desiderata and Theoretical Assumptions

Our desiderata go beyond the typical MAS platforms; in particular, AKIRA is inspired by Minsky's Society of Mind [8], especially with respect to cooperation and coordination of simple Agents. AKIRA borrows many ideas (e.g. micro-level hybridization, coalitions organization, variable speed of the parallel processors) from DUAL [6, 28], that is the main source of inspiration, as well as from Copycat [4] and Boltzmann Machine [26]; however, it is more focused on goal-directness and action-oriented representations, integrating e.g. BDI and Behavior Networks constructs [11].

In recent years many interesting cognitive architectures and models have emerged that try to integrate symbolic and connectionist aspects; moreover, there is an increasing focus on the *architectural* level [20] (how processes and mechanisms are integrated; how the control is distributed between modules and layers; which information is available to the different processes) rather than on single mechanisms.

AKIRA's high-level features can realize the functionalities of many interesting paradigms, ranging from "monolithic" to modular (e.g. the agents are modules that can compete or exchange resources), layered (with different layers performing tasks at different levels of abstraction) and hybrid systems (e.g. DUAL). At the same time, many different architectures, involving symbolic and connectionist concepts, can be modeled and possibly hybridized: in the last chapter we discuss some examples, such as a modular fuzzy system (e.g. each agent represents a rule or a rule set; many agents can fire together), and a hybrid BDI-Behavior Network controller.

General Overview. AKIRA uses hybrid agents (Daemons) having both symbolic and connectionist features; they compete for shared resources and can exchange messages. AKIRA's computational model (sketched in Fig. 1) can be defined as: *a set of concurrent and related processes that incorporate procedures allowing to manipulate shared messages via a Blackboard. Their resources are modifiable at run-time.*

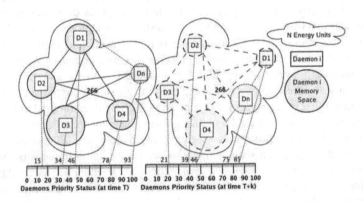

Fig. 1. AKIRA run time: the computational resources of each Agent can change during the computation depending on the energetic dynamics within the system (spreading, tapping)

Resources are not only computational features: they are an index of the *contextual relevance/salience* of an Agent and quantifies the *pressures* it can introduce in the system, e.g. activating or inhibiting some other Agents. Agents can form more complex structures called Coalitions, especially for collaborative and composite tasks. AKIRA furnishes a custom energy model (*AKIRA Energetic Model, AEM*), that involves all the energetic exchanges between the Daemons and with the Energy Pool, allowing developers to exploit the connectionist dynamics of the whole MAS for agent modeling. Its main architectural features are: Hybridism; the Energetic and Physical Work Metaphors; the Locality Principle. As explained, all these features are tools for cognitive modeling: they have to be considered not only as computational choices, but have the status of theoretical assumptions. For this reason, developers can deactivate them (totally or partially). Here we describe the main features.

Hybridism. AKIRA Agents have both a *symbolic* and a *connectionist* component (as in [6]). The symbolic component involves the set of operations an Agent can perform. Depending on the specific simulation, the symbolic operations can range from simple operations to very complex ones (e.g. involving reasoning). However, accordingly to the underlying distributed and MAS approach, complex tasks should be performed by complex-Agents or by Coalitions of cooperating Agents. Hybridism allows developers to perform a continuum of computation styles, ranging from centralized, hierarchical control to distributed, emergent computation. The connectionist component involves the activation level of the Agents as well as their energetic exchanges via the Energetic Network. A major difference exists with many neural-like architectures: the whole system is homeostatic and there are limited resources shared by all the Agents, stored into the Energy Pool; the Agents tap or release energy from it. Spreading energy means losing it (as in the Behavior Networks, [7]).

Two Metaphors. The priority of the Agents is ruled by the **Energetic Metaphor** (see Kokinov, [6]): *greater activation corresponds to a greater computational power*, i.e. speed. Each Agent has an amount of computational resources (energy) that is proportional to its activation level (and is a measure of its relevance in the current context): more active Agents have more priority in their symbolic operations and their energetic exchange, and more frequent access to the Energetic Pool. As a consequence of the Energetic Metaphor, there is intrinsic concurrency in the Pandemonium, even without inhibitory links. Moreover, active Agents introduce a *pressure* over the computation, activating some other related Agents, of forcing the system to follow a certain dynamic. As shown in [6], the pressure mechanism make it possible to model a range of cognitive phenomena, e.g. context and priming effects.

The system implements also a **Physical Work Metaphor**: performing symbolic operations has a cost in energy, that is paid by the performing Agent to the Energetic Pool: this keeps the system conservative. Energetic and Physical Work Metaphors model also the interesting concept of Temperature in Boltzmann Machine [26] as an emergent property of the system. Temperature of the system is represented by the currently used energy, that increases and decreases over time. It shows also how much "structure" and "stability" the system has; and it can be manipulated: for example, it can be raised in order to inhibit Coalitions of Daemons. In Copycat [4] Temperature is proportional to how far it is from a "solution": a hot system is "far from stabilization" and endorses quick-and-dirty calculus; while a cold system is "close to stabilization" and endorses accurate calculus. [3] suggests the interpretation of local dynamics, too;

e.g. high (local) Temperature of a Coalition (and more Agents joining it) can be interpreted as "interesting problem requiring more effort", or *attentional focus*.

3 The Framework

The multi-Agent framework AKIRA is a run-time C++ multithreading environment for building and executing Agents and a web/system development platform to model their behavior and their interaction, as well as for interacting with the environment. AKIRA is conceived and developed using state-of-the-art tools and design: this allows to build applications that are scalable and solid. The AKIRA Macro Language and many automated scripts give facilities and templates for building Agents of different complexity (e.g. reactive, BDI-like). The whole system is written in C++ and integrates many different open source libraries[1], implementing various aspects of the framework. AKIRA is also a Toolkit of resources and algorithms: e.g. many soft computing technologies (Fuzzy Logic and FCMs [13], a versatile Network library) ACLS [23], BDI [12] and Behavior Networks [7] constructs are embedded within the framework as libraries. A solid multithread model ensures parallel computation.

AKIRA follows the Pandemonium metaphor [5]; its components are: the **Pandemonium** (kernel), the **Daemons** (Agents) and the **Coalitions**, the **Blackboard** (XML stream); the **Energy Pool** (an abstraction for the computational resources).

The Pandemonium. The Pandemonium is the system kernel, the main process that instances the threads that are necessary to execute the Daemons (Agents) and that executes all the monitoring and control operations over the single components. Its parameters are configurable at start-up through an XML configuration file; it contains an XML description of: available memory; max number of executable threads; some features for Agents execution (e.g. priority, lifetime, resources); other system properties (garbage collecting, facilities for system and Agents debugging). The Pandemonium Cycle monitors the activity of all the Daemons (including exceptions) and is responsible for many system procedures, e.g. garbage collecting, showing the statistics for Agents, XML stream and system energy. The Pandemonium can also act as a server, collecting and managing requests from many client instances: in this way it is possible to have many kernels implementing multiple societies (using clusters).

The Daemons. The Daemons are the atomic computational elements, each having its own thread and carrying its own code, that are instantiated and executed by the Pandemonium during the system lifetime. They are hybrid [4], having both a *symbolic* component (the carried operation) and a *connectionist* one (activation, tap power). User-defined Agents inherit from an abstract Daemon declaration as well as from many pre-defined prototypes and models. Fig. 2 shows the Agents generation process: the programmer extends some Daemon models; the Agents are dynamically managed

---

[1] Zthread library (threads creation and management); Matrix Template Library (linear algebra); FLIP++/StarFLIP (fuzzy logic); EXPAT (parsing XML); Mersenne Twister (pseudocasual numbers, probability distribution); ACE (networking, patterns for concurrent communication); Doxygen (documentation); Kdevelop (development); Loki (patterns) [27].

by the Pandemonium and start their lifecycle as threads. The semantic imposed by the programmer to the Agents is specified in the *init()* and *execute()* functions. They are called by the framework as part of the run method and used as entry point for each Agent thread. Exiting from run means a regular termination of the current thread with the destruction of everything in its local space.

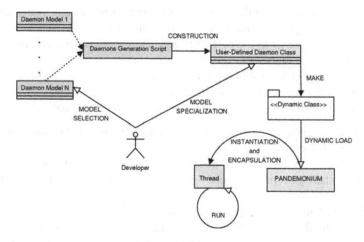

Fig. 2. Agents Generation

The Daemons have some features: a **priority** (set by the programmer), that gives a measure of absolute relevance; a **current activation level** (updated at each cycle), that gives a measure of contingent, contextual relevance; a **tap power**, that sets the access to the concurrent resources (for modeling more or less demanding Agents); a **symbolic operation**, that is the functional body; a **network manager**, manages e.g. spreading activation between the Daemons (but can also load different spreading policies, e.g. the Slipnet [4]: a link is "more conductive" if the concept associated to its label is more active, because it is more contextually relevant). The whole network, associating all the Daemons, is called **Energetic Network**.

The Agents spend all their lifetime within the *run()* functional block, involving:

- *init()* executes some framework initialization steps: setting of Agent state, ID, name into the global names string, timestamp of starting and stop; execution of the user-defined initialization function in the prototype. The Daemon cycle iterates indefinitely until one of the calls to a function cannot be fulfilled. This mechanism is independent from the exception throwing into the blocks and allows a two-layer policy to manage and to monitor failures.

- *tap()* starts the main operative cycle. It allows daemons to tap the Energetic Pool a certain amount of energy as specified in a configuration file (modifiable at run-time). The tapped energy is summed to the base, a-priori Agent energy (*Priority*) and it is used in order to exploit a proportional amount of computational resources. Tap manages the decay operations, too, using a customizable policy.

- *execute()* carries on the symbolic operation of the Agent. It is a wrapper around a function containing externally programmed code (that can be inherited or extended).

- *pay()* is related to the execution of a symbolic operation: each operation has a cost, i.e. the Agent has to release some energy to the Energy Pool.

- *shout()* notifies the results of an *execute()*; it can be used as a monitoring utility for external analysis. Since the notified information resides within the system, it can be even exploited by the other Agents, allowing introspection and meta-level operations, that are highly suitable for cognitive architectures.

- *spread()* gives energy to the neighbors Agents (as defined in a private list); it is the main connectionist function.

- *join()* builds higher-level structures, the **Coalitions**, that will be presented later.

- *reproduce()* duplicates the Agents in order to obtain more resources, as well as to implement evolutionary computation (this topic goes beyond the scope of this paper).

The custom spreading mechanism can be recapitulated as follows: spreading means "giving energy to the other Daemons, and losing it"; tapping is "taking energy from the (limited) Energy Pool"; paying means "giving energy to the Energy Pool". Of course, this mechanism can be replaced by the usual "spreading activation" one [25].

The Coalitions. The Coalitions [6] are communities of cooperating daemons that can be created on-the-fly (differently from *complex-Agents* as previously introduced). Their purpose is to solve together complex, non atomic tasks. For example, in a composite pattern matching problem, each Daemon carries the code for matching a part (e.g. the subject, the sender and the address of an email). Differently from pure connectionist dynamics, Coalitions can cooperate and coordinate exchanging messages (e.g. for interactive tasks requiring explicit coordination). Coalitions arise and die in a dynamic way during computation when Daemons join and leave them.

There are two kinds of Coalitions: *Bands* and *Hordes*.

- *Bands*. Bands arise as the result of an auto-organization process, e.g. in order to find help for a non atomic problem, in which its single symbolic operation is a non sufficient part. Their semantic is mainly driven from similarity and proximity (via the Link List) and concomitant activation (only active Daemons can join Coalitions).

- *Hordes*. Hordes arise in a more top-down way and have a more structured, hierarchical shape. Special purpose Demons, **Archons**, try to recruit other Daemons, spreading them some energy, and take advantage of the symbolic operations they perform. Archons are not directly executers: they carry on a non-atomic *Structure*, in which specific *Roles* to be played by appropriate Demons. For example, the structure can be a set of successive operations to be performed (as in a Plan); in this case the Archon will activate (e.g. sequentially) a number of Daemons (e.g. actions) that play the role of performers of those operations. While in Bands the aggregation starts to build a structure in a "blind" way, in Hordes the prototypal skeleton of the structure is carried on by Archons. Another example is a complex pattern matching operation for recognizing the words "THE CAT" even if "H" and "A" are graphically identical (as shown in Fig. 3). This operation can be performed using a "layered-like" structure: the Daemons in the first "layer" (the clouds) are feature-recognizers; but at this level the "H" and the "A" can not be discriminated. The Daemons at the second "layer" are Archons, carrying on the prototypical structure of a "H" and of a "A"; they are linked "at a different level" that involves word-level and syntactic knowledge, i.e. the words "THE" and "CAT", represented by the Archons in the third layer.

Bands and Hordes can of course interplay and shift within the same system: for example, the pattern matching process can start in a bottom-up way, e.g. some feature-recognizer Daemons arise and start to form a Band (the clouds in Fig. 3); their operations can be noticed by letter-recognizer Archons, that continue the operation by sending activation the other (possibly) relevant Daemons that could be not jet active.

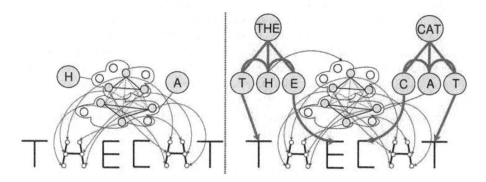

Fig. 3. Bands and Hordes for a complex pattern matching operation

Message Passing. In AKIRA three message passing mechanisms are available.

- The *Blackboard* (XML Stream) is a shared data structure divided into blocks containing AXL (AKIRA XML Language, a custom KQML-like data exchange language) packets, where the messages are concurrently written and read.
- *AkiraGenericObjectFactory* allows developers to create, set, get and destroy on the fly shared objects of any kind; it is the slower mechanism.
- *AkiraGlobalVariableFactory* is limited to shared variables of scalar type.

All the mechanisms have a *templatized mutual exclusion* policy to guarantee thread safe and consistent access to all data.

Energy Pool and Energy Dynamics. Energy is not a component; however, its dynamics are central in explaining the functionalities of the system. A global variable, energy, shared by all the Agents, gives an upper bound of the available computational resources. The access is regulated by a private templatized mutual exclusion policy. The energetic level of the system, summing up all the priorities for all the Agents, their tap power and the Pandemonium energy, gives an upper bound to the possible activation sequences of the single threads during the AKIRA lifetime. For each Agent, more energy means more resources (e.g. more computational time). Energy is spread by the Agents through the Energetic Network, driving the sequences of activation of the threads, without violating the energy conservation law.

Summary of the Operations. The Pandemonium is the "father process": in the start-up phase it identifies the Agents to load (accordingly to the constraints in the initial configuration); during the execution it monitors the content of the Stream and of the single Agents. The Agents, whose resources may change dynamically, execute the code in their body if it fits the situation (like productions); they run for a given (even indefinite) amount of time.

In synthesis, this is the procedural scheme of an execution of AKIRA: (1) **Start**: static structures are loaded and objects are initialized for the Pandemonium and the XML Stream. (2) **System Initialization**. Configuration of the system parameters and searching for external Agent classes. (3) **Instantiation and initialization of the Agents**: creation and storing of the Agent instances associated to the external classes and setting of their parameters. (4) **Agent execution**: each Agent is executed by the Pandemonium with an associated thread. (5) **Control cycle**: (a) Analysis of the state of the Agents and garbage collection of non executing Agents; (b) Analysis of XML Stream and garbage collection of old packets. (c) Print Agent state. (d) Print the content of the XML Stream. (e) Print the available energy of the system. (f) Suspend for a given amount of milliseconds (depending from the priority).

Fig. 4 sketches the interplay between the main components: the Pandemonium, the Daemons and the Blackboard. Daemons share a limited amount of energy, too.

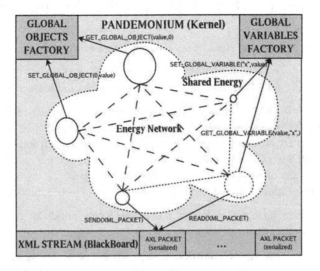

Fig. 4. schema of interaction of the components in AKIRA

All the operations follow the *Locality Principle*: Every interaction between Agents is implemented by a local rule; the only global operation is the energetic exchange with the Energy Pool. The medium for all the operations is the Blackboard, that is however only a functional abstraction and not a serial bottleneck. The connectionist side of AKIRA endorses many emergent properties, implemented in a MAS perspective, allowing computational units to be complex and autonomous. Moreover, simple local interactions between Agents can lead to complex group dynamics (e.g. *Coalitions*); as shown in [14], many coordination and cooperation tasks can be achieved in a distributed way, without centralized control.

4 A Society of Functions

A strong motivation behind AKIRA is having a framework solid enough to build real-world and scalable applications, and allowing agent designers to have a number of cognitive functions implemented by different underlying mechanisms. While in AI and agent technology there are a number of interesting problems that are well understood and solved using a specific, successful mechanism, it is at least questionable if there is (at the moment) a mechanism that is scalable enough to be successfully applied to a number of diverse applications. At the same time, an interesting feature of MAS is that different agents can solve the same (or related) problems, or even coordinate and collaborate, without implementing the same function or mechanism; moreover, they are not even supposed to know or to understand each one the functioning of the others. It is the core idea of the Society of Mind by Minsky (1986): *a number of narrow-minded 'specialists' that learn how to coordinate and to exploit one another in order to fulfill a selfish or a common goal.*

AKIRA is also a *Toolkit* of resources: the AKIRA Macro Language contains many of state-of-the-art facilities, algorithms and mechanisms (including fuzzy calculus, neural networks, belief networks, behavior networks, anticipatory classifiers, BDI constructs, genetic algorithms, etc). At the same time, we are building interfaces with a graphic engine, developing a number of architectural facilities (web based programming interfaces, integrated analysis tools), and writing an interface to a descriptive ontology (DOLCE, [15]).

In the *cognitive science perspective*, we use AKIRA as an "experimental laboratory", for exploring a range of models and mechanisms, in order to see how they interact, collaborate and exploit one other in order to manifest higher-level cognition. In the *agent design perspective*, AKIRA allows developers to implement and test many agent models, in order to exploit the best "design level" for each kind of application and to focus on the architectural level.

Designing at the Functional Level. We made an effort to keep separate in AKIRA the functional and the implementation levels. As cognitive scientists, we are interested in designing and modeling *functions* rather than *mechanisms*; this means that a designer should be allowed to build and test his architectural model e.g. goal oriented behavior, independently from its physical and computational realization.

For this reason, AKIRA allows to design agents both as *simple* and *complex*; while for a simple agent a single Daemon is used, complex agents are modeled using many Daemons that compete for resources, collaborate and exchange messages and activation. This means that, whichever it is the selected agent model, it is possible to realize the same logic at different level of detail and with different computational resources, involving one or more interacting daemons. For example, a BDI architecture can be implemented either in the body of a single daemon (e.g. with a central interpreter) or as the emergent activity of a set of related daemons (e.g. each representing a goal, a belief, etc). It is possible for the programmer to design the abstract architecture, and then choose the level of representation granularity for the agents (and even for the environment). At the same time, AKIRA provides a set of computational facilities and mechanisms (e.g. fuzzy logic, neural networks) that can potentially realize the same functions. The programmer can make experiments with the different tools and try to implement its functions in different ways.

It is worth pointing out that these choices do not only involve computational power and resources: in some cases the mechanism and functional levels are not totally independent, so the behavior and the capabilities of the same architectural scheme may vary according to the underlying model. Thus, a choice involves a theoretical commitment towards different kinds of models (serial vs. parallel and distributed) and mechanisms (e.g. a certain kind of logic). We think that using AKIRA as an experimental laboratory can help to answer to some interesting question such as: which mechanisms can realize the same functions? Which functions strongly rely upon their physical or computational infrastructure? Which ones are untranslatable?[2]

Exploiting the Coalitions. The concept of Coalition is abstract and their behavior is highly customizable: the Coalitions can be realized and exploited in different ways, depending from the application. The Coalitions are one of the more interesting features, still under-explored, that can be exploited by MAS in order e.g. to model collective behavior at different level of complexity and with different level of computational and cognitive involvement. There can be Coalitions that arise for bottom-up dynamics of the system, allowing to characterize e.g. emergent phenomena and auto-organization. Many biologically inspired systems can be simulated in this way (at the same time, their dynamics can provide a valuable source of inspiration for realizing complex tasks with few resources). Coalitions can also be used for more structured, hierarchical and organized tasks: they can arise in a more top-down oriented way, because of the *pressure* of specific Daemons (Archons), specialized in recruiting other Daemons in order to build complex structures or procedures. Of course, there is a functional continuum and an interplay between all these modalities.

While Daemons can take advantage of many forms of pure connectionist learning (e.g. involving the Link List), the Coalitions allow also to perform symbolic learning. For example, a new Archon can be created by to bottom-up pressures (e.g. if a Band

---

[2] The Functional Level Design approach is, in a certain sense, the opposite of the Unified Theory of Cognition (e.g. Soar, [21]), where it is not clear if the *mechanism* (e.g. production rules) or the *function* it realizes is assumed to be universal (but we are induced to assume that the mechanisms are). A reduction of the theory to the implementation underlies also many works in the neural network field, where the cognitive models are directly mapped into the NN mechanisms; in other terms, the NNs as such are presented as the model, not the information processing structure that they realize. In this way, the NN computational features (e.g. their dynamics, decay and fire rate) become implicitly part of the models and constrain the theory, sometimes acritically. Moreover, it is impossible to conceive interesting questions such as: Which are the influences (and side effects) of the specific mechanisms over the functions and models? Does it heuristically suggest something new? Is there a certain feature necessary for a specific function? Or are the cognitive functions mechanism-independent? Designing at the level of functions means choosing the better mechanism and level of detail of implement them.

shows persistence) or by top-down motivational pressures (e.g. via analogy). The prototypical structure of an Archon can also be modified e.g. learning new exemplars of situations, as in Case Based Reasoning. There are many facilities at the framework level for implementing, testing, and using Bands and Hordes; their implementation (and their semantic) is application-specific. However, for many typical and interesting MAS tasks (e.g. involving coordination and cooperation) it is extremely useful to have elaborated tools to design at a non-atomic level, i.e. at the Coalition level.

Resources, Urgency and Pressures. A Demon or a Coalition can act (i.e. *execute*) only if it has sufficient energy, because each symbolic operation has a cost. The cost can be seen as inverse to *urgency* of the behavior: less cost means more easily activated. So, urgent behaviors, like stimulus-response behaviors (as well as *alarms* [20]), can be represented with very low-cost operations; more complex cognitive operations are slower: they need to recruit a lot of energy, or exploit operations by other Daemons, or wait for one or more *join*.

As a consequence of system dynamics, each Daemon introduces a *pressure* over the computation in virtue of its presence. The system shows also implicit, contextual forces and pressures (e.g. set points) that may lead to Coalitions.

Daemons and Coalitions introduce contextual pressures in many ways: perceptual, goal-driven, cultural, conceptual, memory contexts are among those possible. As an effect of the Archons work, Demons that are somewhat related to the contexts are able to recruit more energy; this is true even if they are not able to join the Horde. But this is also true of Bands, where the pre-existing link topology is assumed to embed, in an implicit way, a "similarity" or "categorical" semantic (much like SOMs, [24]).

4.1 Simulation Models in AKIRA

The consequence of hybridism and of the other global properties is that AKIRA allows to implement an *homeostatic* MAS, endorsing *concurrence for the resources* between the Agents. What kind of Agent Societies can be implemented in AKIRA?

We define an *AKIRA Agent Society* as a set of Agents working under a common Energetic Pool and competing for its limited resources; an *AKIRA Agent* is the unity which has a single, concurrent access to the Pool, proportionally to its activation/salience (absolute and contextual). All the energetic exchanges are local between the Agents; all the symbolic operations have a cost in energy. The system is conservative and it allows emergent phenomena.

AKIRA is open enough to allow many kinds of agent models and societies; it fits the general MABS requirements as individuated in the Introduction (perform simulation with Agents having different level of complexity, interacting within dynamic environments, etc.). However, some of its features (e.g. the Energy Pool), are especially conceived for a certain kind of Agent design, constraining the kind of simulations it can endorse. For example, in many social simulations, resources management involves manipulating some "environmental variables"; in AKIRA it can be delegated to the built-in energetic dynamics, assuming that the underlying metaphors are accepted. There is always an influence of the architectural mechanisms over the models they support; we have tried to make all the assumptions and implementation choices transparent enough to be adopted or deactivated.

Here we shortly present some social and socio-cognitive simulations we have performed; even if their complete discussion is beyond the scope of the present paper, describing them may help to understand the unique features of AKIRA.

4.2 AKIRA at Work

The **Vampire Bats Scenario** [19] involves very simple Agents. It is an (ecologically plausible) predator-prey variant, that includes altruism and reciprocity: bats can give some food to unnourished ones, increasing kinship survival rate. Bats are very simple Agents, implemented with a single Daemon. The internal motivational state of the bats is modeled with a Fuzzy Cognitive Map, while their behavior is governed by fuzzy production rules. The environment is a single Daemon, too, responsible for some global variables such as simulation time (clock), night and day succession, and food resources. For this simulation we deactivated the energetic network (because bats do not exchange energy). There is also a roost formation and migration mechanism implemented using the Coalitions: bats exchange food only within a roost. We studied the survival dynamics as a function of roost number and dimensions, performing both turn-based and real-time, parallel simulations (for model-to-model comparison). In real-time, we modeled the resources (the available food) as the energy into the Energy Pool; the bats were implemented as Daemons concurring in parallel for limited resources, and each operation (move, chase) had a certain energetic cost. The compared results show that it is possible to perform this kind of simulations exploiting the built-in mechanisms of AKIRA, without extra variables.

We have also implemented a number of other **social scenarios** with Agents having more complex cognitive features (e.g. contract nets; societies with reputation and trust dynamics [11]). Even in this case each Agent is a single Daemon; however, each Agent is able to perform complex decisions, building representations of other Agents' features (e.g. ability, reputation, reliability). These simulations exploit many soft computing facilities embedded in the framework, e.g. fuzzy logic. Moreover, even if the Agents are self-motivated, they can form Coalitions in order to better achieve their own goals (e.g. purchase at a better price in a market, collaborating in a task).

The **Watchdog Agent Scenario** involves many interacting Daemons for each Agent (e.g. one for each BDI and behavior network component). The Watchdog, that has to patrol a house, is composed of many Daemons, including: a Norm: *stay always close to the house*, and some Goals: #1 *walk around the house*; #2 *bark if you see an intruder*; #3 *chase and follow the intruders*. Goals inhibit each other via the Energetic Network. The architecture is hybrid, using BDI, ACLS, Behavior Networks.

In order to fulfill goal #1 respecting the Norm, the Watchdog follows circular trajectories around the house, standing always close (in fuzzy terms) to it. When it individuates as intruder, in order to fulfill *goal #2*, the Watchdog will bark; if the intruder tries to fly out, in order to fulfill *goal #3* the Watchdog has a pressure to follow it. In this case the *Norm* and *goal #3* are two contrasting pressures: the first to stay close to the house, the second to leave the house.

The Watchdog trajectory results from a mix of those factors; more, the internal dynamic of the system will follow some built-in rules for Goals and Norms: the goal becomes stronger the closer it is from its realization; the norm becomes stronger the farther it is from its realization; both become stronger as the Watchdog follows the intruder. So, assuming a slightly higher priority for the Norm, the Watchdog will follow the intruder, until: either the dog reaches him, or it goes too far from the house and the pressure of the Norm becomes stronger.

The behavior of the Watchdog simply results from diverging pressures: the trajectory as well as the exact point where it comes back home are not pre-calculated. However, the effect can be magnified by a symbolic operation; e.g. after a while (when its clause is far from realization) the Norm can activate another goal: #4 *come back to the house*. Or there can be the contribute of explicitly planned activities: a Goal can activate a Plan involving a rigid "sentinel routine" e.g. follow a certain trajectory that includes each corner, bark each ten minutes, etc.

The Watchdog behavior thus emerges from the *interplay between top-down and bottom-up components and pressures*. It can start as a reactive, stimuli-driven action, be modulated by contextual pressures, activate a Goal and shift to a proactive, top-down control sequence regulated by a Plan; or it can go the other way.

Exploiting Opportunities. Having many concurrent processes active at the same time means that some of them can exploit the opportunities offered by the others, i.e. profit of operations performed for other purposes. This is the other side of the concept of pressure: the activity of a Daemon or Coalition not only influences the others and "demands attention", but can also be exploited by others.

For example, in goal processing this means that the system needs not to plan everything in advance; at the contrary, a process can exploit the (results of the) actions performed by other processes for their own purposes. E.g. in the Watchdog example, the goal "eating" can exploit the "movement towards the house" produced by the Norm (if some food is in the house). In a slightly different case, two processes can both contribute to the activation of the same action for different purposes.

This property accounts also for many evidences in cross-modal interaction, e.g. the results of a visual process can be exploited to prime and facilitate an acoustic one. This is impossible in a modular architecture where each sub-process is realized in a completely independent way and is not influenced by the pressures of the others.

Current Projects. Currently we are exploiting AKIRA for many other cognitive modeling tasks, implementing Goal Oriented, BDI-like constructs in a parallel and dynamic fashion, and extending the formalism to many fields in which ISTC-CNR is concerned (e.g. expectations, epistemic actions, constructive perception, monitoring and control) [9]. In the frame of the MindRACES European Funded Project, focused on *anticipation*, we are exploiting AKIRA in order to model and implement within the same framework *a number of different forms and functions* of anticipations, situated at *different levels of cognitive complexity*, and possibly exploiting and comparing different mechanisms.

For example, the Watchdog will be enhanced with the capability of performing *epistemic actions* [22], e.g. "look at the world", "control if"; in this way it is proactively driven by its expectations instead of merely reacting to input data.

5 Conclusions

We have described AKIRA, a MAS platform for cognitive and socio-cognitive agent modeling and implementation. AKIRA couples the MAS and the Pandemonium metaphors. The Pandemonium is the system kernel, that runs and monitors the agents. The main actors are the Daemons, that communicate through a Blackboard and exchange energy through an Energy Network and a common Energy Pool. Daemons are hybrid: they have both a symbolic and a connectionist side. With respect to their **symbolic side**, they can: *execute* the symbolic action they carry on, that can be performed if the contextual conditions are met (like productions) and if the energy is sufficient; *shout*, notifying their current activity and status to the other Daemons via the Blackboard. With respect to their **connectionist side**, they can *tap* energy from the Energy Pool; *spread*, giving it to linked Daemons; *release* it to the Energy Pool; *join* other agents in order to form complex structures called Coalitions. Using the Daemons it is possible to implement *simple* and *complex* Agents, i.e. Agents having one or more Daemons as constituents. In order to allow rapid prototyping of Agents at different level of complexity (reactive, deliberative, layered) we have included in the framework the *AKIRA Macro Language*, that furnishes a number of facilities and tools for agent programming, including many soft computing technologies (e.g. Fuzzy Logic and FCMs) and many BDI constructs (beliefs, goals, desires).

We have also described some theoretical constraints that make AKIRA unique with respect to the MAS: mainly Hybridism of the Agents (symbolic and connectionist) and a set of System Properties (the Energetic and Physical Work Metaphors; the Locality Principle). We have discussed what kinds of simulations are better performed using AKIRA, and presented some examples.

AKIRA is in *pre-alpha 2_stable version* at SourceForge [16], running under Linux platforms, both in serial and parallel processors. With respect to some well known simulative [17, 18] and Agent [10] platforms, AKIRA still lacks some facilities, such as user-friendly programming and monitoring interfaces; support for data analysis. However, it has proven to be suitable both for MABS and as an Agent design tool.

In the *technological perspective* AKIRA shows state-of-the-art design and implementation that makes it highly suitable for software engineering; this includes a solid thread model (based on Zthread API and POSIX underlying features) that allows real parallel computation. The Pandemonium can also act as a server object, accepting incoming connections from clients via AXP Protocol. Many Pandemonium instances can link each other in a cluster-like way.

In the *application-specific perspective*, some high-level features (e.g. Hybridism, the Energetic Metaphor) allow developers to model complex system dynamics, such as Coalition formation, interplay between top-down control and bottom-up, emerging behavior. It does not imply committing to a specific Agent model, because many models can be implemented (and all the built-in features can be deactivated). However, the strengths of the framework better emerge using a certain kind of Agent design, i.e. exploiting the Coalitions and the Energetic Network for concurrency and cooperation; using the Energy Pool for managing limited resources.

Acknowledgements

The SourceForge open-source community [16] is an invaluable resource for this project, providing state-of-the-art methodologies and development, insightful discussion and strong motivation. This work is supported by the EU project "MindRACES: from Reactive to Anticipatory Cognitive Embodied Systems".

References

1. Castelfranchi, C; Falcone, R., Principles of trust for MAS: Cognitive anatomy, social importance, and quantification. *Proceedings of the Third International Conference on Multi-Agent Systems*, Paris 1998.
2. Conte, R. Emergent (Info)Institutions, *Cognitive Systems Research*, , 2001, Vol. 2, Iss. 2
3. B. J. Baars. *A Cognitive Theory of Consciousness*. Cambridge University Press, Cambridge, MA, (1988).
4. D. R. Hofstadter and FARG, *Fluid Concepts and Creative Analogies: Computer Models of the Fundamental Mechanisms of Thought*, Basic Books, New York, (1995).
5. J. V. Jackson, Idea for a Mind. *Siggart Newsettler*, 181:23-26, (1987).
6. B. N. Kokinov, The context-sensitive cognitive architecture DUAL, in *Proceedings of the Sixteenth Annual Conference of the Cognitive Science Society*, Lawrence Erlbaum Associates, (1994).
7. P. Maes, Situated Agents Can Have Goals. *Robotics and Autonomous Systems*, 6 (1990).
8. M. Minsky. *The Society of Mind*. Simon and Schuster, New York, (1986).
9. G. Pezzulo, E. Lorini. G. Calvi. How do I Know how much I don't Know? A cognitive approach about Uncertainty and Ignorance. *Proceedings of COGSCI 2004*.
10. F. Bellifemine, A. Poggi, G. Rimassa. JADE – A FIPA-compliant Agent framework. *CSELT internal technical report*. 1999
11. R. Falcone, G. Pezzulo, C. Castelfranchi, G. Calvi. Why a cognitive trustier performs better: Simulating trust-based Contract Nets. *Proceedings of AAMAS 2004*.
12. A. Rao, M. Georgeff, BDI Agents from Theory to Practice, *Tech. Note 56, AAII*,1995.
13. Kosko, B. Fuzzy cognitive maps. International Journal of Man-Machine Studies. 1986 (24)
14. Mataric, M. J. (1995). Designing and understanding adaptive group behavior. *Adaptive Behavior*, 4(1).
15. Gangemi A., Guarino N., Masolo C., Oltramari, A., Schneider L. Sweetening Ontologies with DOLCE. *Proceedings of EKAW 2002*. Siguenza, Spain.
16. http://www.akira-project.org/ and http://sourceforge.net/projects/a-k-i-r-a/
17. Minar N., Murkhart R., Langton C., Askenazi M., *The Swarm Simulation System: A Toolkit for Building Multi-Agent Simulations*, http://www.santafe.edu/projects/swarm/, 1996.
18. CORMAS - *User Guide* http://cormas.cirad.fr, 2001
19. G. di Tosto, M. Paolucci, R. Conte (2003) Altruism Among Simple and Smart Vampires *Proceedings of ABM2003*
20. Sloman A. *What sort of architecture is required for a human-like agent?* In M. Wooldridge. A. Rao (eds). Foundations of Rational Agency, Kluwer Academic Publishers, 1999.
21. Rosenbloom, P. S., Laird, J. E. & Newell, A. (1992) *The Soar Papers: Research on Integrated Intelligence. Volumes 1 and 2*. Cambridge, MA: MIT Press.
22. Kirsh, D. & Maglio, P. (1994). On distinguishing epistemic from pragmatic action. Cognitive Science, 18, 513–549.

23. Butz, Martin V. (2002) *Anticipatory Learning classifier systems*. Boston, MA. Kluwer Academic Publishers.
24. Kohonen, T., *Self-Organizating Maps*, New York : Springer-Verlag, 1997.
25. Collins, AM, and EF Loftus (1975) *A spreading-activation theory of semantic processing*, Psychological Review 82, 407-428.
26. McClelland, J. L. & Rumelhart, D. E. (1988). *Explorations in Paralell Distributed Processing: A Handbook of Modles, Programs and Exercises*. MIT Press, Cambridge, MA.
27. A. Alexandrescu (2001), *Modern C++ Design: Generic Programming and Design Patterns Applied*. Addison Wesley Professional
28. Kokinov, B. (1997). Micro-level hybridization in the cognitive architecture DUAL. In R. Sun & F. Alexander (Eds.), *Connectionist-symbolic integration: From unified to hybrid approaches* (pp. 197-208). Hilsdale, NJ: Lawrence Erlbaum Associates.

Work-Environment Analysis: Environment Centric Multi-agent Simulation for Design of Socio-technical Systems

Anuj P. Shah and Amy R. Pritchett

School of Industrial and Systems Engineering,
Georgia Institute of Technology,
Atlanta, GA – 30332, USA
{ashah, amy.pritchett}@isye.gatech.edu

Abstract. This paper presents a multi-agent based simulation framework for cognitive systems engineering of socio-technical systems. Comprehensive design analysis of socio-technical systems requires modeling of various aspects of the work environment and of behavior and performance of humans. This framework provides a distinct focus on the work-environment, specifying it as a coherent collection of declarative models spanning its multiple aspects. Compared to traditional methods in cognitive systems engineering, aggregating multiple aspects allows greater detail and scale in modeling socio-technical systems. This also addresses design issues that cut across individual aspects, thus enabling a comprehensive what-if analysis of the system. Humans are computationally modeled as proactive and interactive agents operating within their work-environment. Emergent behavior of the system in response to design changes in both humans and their work environment can thus be simulated. The framework is illustrated through an example in air traffic control. The framework can be applied to problems in enterprise re-engineering, organizational structuring, etc.

1 Introduction

When designing large-scale systems that constitute humans, technologies, processes, organizational structures, etc. one is concerned with how changes in each of these elements may affect the system's overall performance. Whether these changes are novel or whether they are incremental, a priori analysis of their system level effect is very important, especially when the changes are too expensive to alter after implementation.

As the scale and/or the number of design variables increase, the complexity of behaviors within the system can rise to levels unmanageable with current systems engineering models. Current models are often specified through the use of natural language, rendering them impracticable for complex system analysis. Concrete representations for computer simulations are needed for complex system analysis.

This paper describes a framework for a priori analysis of system design variables such as technology, processes, etc. The framework is computational for design

P. Davidsson et al. (Eds.): MABS 2004, LNAI 3415, pp. 65–77, 2005.

analysis of large-scale and complex systems such as transportation systems, military organizations, and enterprises.

The following section introduces the research issues in complex socio-technical systems design. After summarizing the research issues, the Work-Environment Analysis framework is described with respect to these research issues. The framework is then illustrated through an example in control-procedures analysis for airport approach control. The paper concludes with a discussion of the contributions of this framework, with respect to the field of multi-agent simulations and its applications.

2 Design Issues in Socio-technical Systems

Socio-technical systems (STS) are the building blocks of everyday life. The term has traditionally been used to identify technology in its social settings, but now it identifies systems that comprise people, technologies, physical surroundings, processes and information [1]. Examples include transportation systems, military organizations and corporate enterprises.

STS function by local action and interaction of entities that make up the system. These entities include the humans working in the system and their environment that affects and shapes their cognitive choices and the output of their work [1, 2]. The macro-level behavior and thus the measured performance of the STS *emerge* from the local (micro-level) actions of humans and their interactions with each other and with their work-environment. *Emergent* is defined here as a system property in which system behaviors at a higher level of abstraction are caused by behaviors at a lower level of abstraction which could not be predicted or estimated at that lower level. Thus one could model the entities that make up the system and simulate them to see what system behavior comes out or emerges.

Thus, firstly, due to its emergent properties a STS is essentially a multi-agent system, the system level behavior of which emerges from the micro-level properties and behaviors of the proactive and interactive humans (modeled as agents) and their interaction with the environmental elements [3, 4]. Secondly, the design variables in the system (i.e., the aspects of the system that can typically be specifically changed) are the humans, elements of the work-environment, and their interrelationships; all affect system performance in a manner that can be difficult to estimate *a priori*. The second observation calls for a *comprehensive approach* to system design and analysis, i.e. an approach that aggregates design factors both in humans and in their work-environment, thus considering all system design variables. Work-Environment Analysis (WEA) framework presented in this paper evolves from these basic observations.

In the following the two design variables, i.e. the work-environment and human behavior within STS are further discussed to build up the motivation for the WEA framework. Technological issues in engineering large-scale STS through simulation are then discussed.

2.1 The Work-Environment

All objects, technologies, processes and information that surround the people in STS make up their environment. These environmental elements relate to each other in context of the work being performed in the system in several ways. For example, objects and technologies that can be employed for accomplishing system tasks are related through task-resource relations. Similarly there could be aspects that identify with processes, organizational structure, and data and information.

These aspects can be modeled to approximate the environment in a manner suitable for STS design analysis. However, by limiting the models of the work-environment to a selected few of its specific aspects, one cannot answer the questions that go beyond those aspects and that range through multiple aspects. The interplay between the aspects may lead to emergence of dynamics that may not be manifested through exclusive examination of each one of them. For example, STS designers may be faced with questions such as:

1. What mix of technologies and processes provides a more efficient and safer system?,
2. How well do the system technologies and processes generalize to a different mix of workers or organizational setups?
3. What aspects of the environment should the agents be trained on to meet system objectives?,
4. Which design of the work-environment exhibits universality in system performance for given range of variation in human behavior?, etc.

The first question requires the designer to consider not just a mapping of tasks to the resources in the system, but also the processes that apply on the agents in the system. Similarly the second question adds an aspect of organizational setup that may be specified in terms of roles and responsibilities.

Many cognitive systems engineering approaches are intended to examine the third and fourth questions. However, so far they have considered at most one or two aspects of the work environment. For example, Cognitive Work Analysis (CWA) works with functional and parts-whole hierarchies in the work-environment [1]. These aspects can primarily be employed to solve design problems that efficiently and effectively employ system resources in individual, team or organizational settings. Cognitive Triad (CT) employs goal structures and task-resource maps in the work-environment for design of decision support systems for strategy and tactics selection [5]. CWA or CT cannot answer questions one and two. To answer those questions, they require consideration of other aspects such as the organizational aspect along with the aspects they support. Moreover, they are limited in detail for answering the third and the fourth questions.

Detailed STS design requires a mix of environmental aspects. A comprehensive approach that can aggregate multiple aspects including resources, functions, goals and processes is therefore needed. The designer should be able to pick and apply multiple aspects of the work-environment to address the design question(s) being answered. But, it is important to maintain a parsimonious approach that considers only the necessary and sufficient aspects.

2.2 The Agents

As previously discussed, the behavior and performance of a STS emerges from the interaction of its agents with each other and with its work-environment. Naturally, when the work-environment changes, human behaviors need to be tailored appropriately. For example, when automating manual work, workers need to be trained to use this new technology using new processes. Thus, for each new design of an STS's work-environment, agents tuned to these changes have to be generated to comprehensively model the important system behaviors.

2.3 Engineering Issues

Design complexity becomes significant when considering multiple aspects of the work-environment, with corresponding agent designs. This complexity tends to scale non-linearly with the design's scope and detail, quickly becoming unmanageable. Since most of the design problems concern large and complex real world systems, there is an unmet need for a design framework that manages this complexity while providing the insight required by designers.

A structure preserving modeling framework is one method to help reduce design complexity. System properties are not understood at the system level, but design changes typically only occur at the component level. When the structure of system components is preserved during modeling, design and analysis activities, designers can easily translate between the real world and the design models. Preserving the structure also helps keep the interactions between components realistic.

A structure-preserving framework also aids when examining emergent phenomenon within the STS. By maintaining the correct component-level structures, their impact on system-wide emergent behavior can be specifically examined, even to the point of investigating the causality of decrements to system performance.

The designers should not have to create the implementation for design analysis of each design intervention/iteration. The design should be used to generate all executables to be directly run in the design analysis engine. Thus the framework should be design driven, i.e. each design variable should be concretely specified to the level that it can be directly employed for the analysis activity.

Appropriate data capture and measurement utilities should also be created to capture data for design evaluation, irrespective of the domain or scope of the design problem. This should be done at the level of abstraction of the design components (micro-level) as well as that of the system (macro-level) for offline or online analysis.

Other major technological issue is that the framework should fit to any domain and to systems of any scale. Thus the system should be re-configurable to different kind of work-environment and agent models. These designs should not be limited in number of components that can be used, or size of components that can be used.

3 Work-Environment Analysis

The Work-Environment Analysis (WEA) framework employs multi-agent simulations to perform what-if analysis on system models. It enables modeling of multiple useful aspects of the work environment into one coherent environment model, thus significantly enhancing the scope of system analysis and design. It employs computational agent models together with declarative and extensible models of the work environment in large-scale simulations for what-if analysis of socio-technical systems. It also enables designers to directly manipulate the design variables without having to recode the agents with each change.

Uniqueness of WEA lies in its feature that it can be dynamically configured to accommodate any number of different cognitive environment models that can be concretely specified in a declarative fashion with the use of XML. But each declarative element should have a (set of) corresponding executable component(s) that can be employed to configure agents with environment-aware capabilities.

Figure 1 shows the WEA framework. The framework follows a very simple simulation and analysis architecture, where the inputs to the simulation engine include models of the agents and the work environment, operating scenarios and the system performance metrics. The simulation engine runs the multi-agent simulations and logs it's the system performance that emerges. The following details the components of the WEA simulation framework.

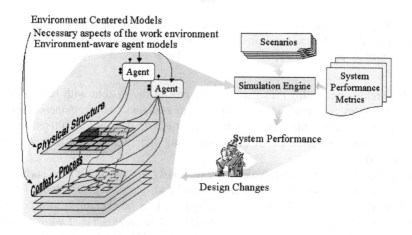

Fig. 1. The Work-Environment Analysis simulation framework

3.1 Models of the Work-Environment

A designer starts with modeling of necessary aspects of the work-environment. Each aspect relates the elements in the work-environment through specific work oriented relations. Since there could be many work-oriented relations between the same environmental elements, the same environmental element could be represented in multiple environmental aspects. Therefore, it was chosen to model each element as a *container* of declarative components, each of which represents it in different environmental aspects. Each environmental aspect is represented as a connected graph, each node of which is one of these declarative components of the elements (refer Fig. 2). The arcs of the graph specify the work-oriented relations. That is, each environmental aspect is defined by a set of semantic relationships that link the environmental elements into an environmental aspect. The *container* model of the environmental element also enables addressing coherency between the different aspects by maintaining each element's identity is maintained across the aspects.

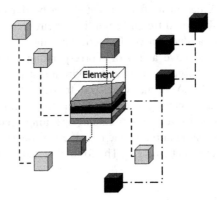

Fig. 2. Modeling environmental elements and multiple aspects of the work-environment. The figure shows three different environmental aspects (represented by connected nodes), with environmental elements (box shaped nodes) connected in graphs

This architecture helps preserve the structure of the system with respect to the design variables, i.e. each element in the environment, its components and the work-oriented relationships. Each of these entities can possibly be altered towards system level design, and each of them represents a tangible design intervention in the real world.

The fidelity of these models depends, obviously, on the fidelity of each of the components, the fidelity of the container model and that of a graph model for each aspect. In theory, these principles can be applied to a number of known cognitive engineering models such as the Abstraction Hierarchy [1], Task-Knowledge Structures [6, 7], Functional Abstraction Hierarchy [5], Goal Trees [5], Context-Process Structures (devised in this research) etc. Context-Process structure will be explained as an example, with the example design problem in section §4.

3.2 Environment-Aware Agent Generation

Once the environmental aspects have been modeled, agents aware of this environment are modeled. These models are used to automatically generate agents for the simulation. The basis of these generative models could be an organizational chart, or an attempt to fit the available agents (defined by a set of capabilities and objectives) to the current work-environment. Currently, the designer manually performs this modeling activity. This activity could later be automated to search through a space of agent configurations to derive the best fit of agents to the environment.

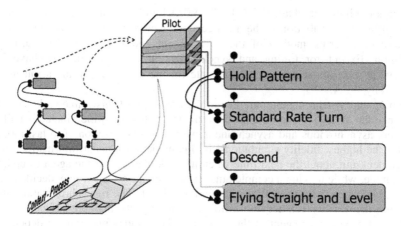

Fig. 3. Generating a Pilot agent based on environmental processes that it needs to be aware of, that in turn dictate the agent's required capabilities

Figure 3 shows the architecture for generating agents. This architecture is closely related to the environment models. An executable component is attached to each of the declarative components of the environmental elements. These executable components can be directly plugged into the agent models to construct an agent capable of working with the respective environmental elements. Each of these components can work with all related components in the environmental aspect. That is, the interface definitions of these components are specified to the level of detail needed for operation with other elements in the work environment.

Consistency and completeness checks can be done at the model level. The agent capability-components can build up on each other. These components can have any internal architecture and thus the agent model is open to any cognitive architecture. Each component perceives the environment and acts upon the environment through standard interfaces of variables and methods. In the constructed agent model these methods and variables are all added to the context of the agent. The agent's internals are thus hidden from all external components. That is, an agent interacting with the other agent does not have to know the internal capability it is dealing with. This makes the architecture dynamically re-configurable and highly extensible.

3.3 Other Architectural Issues

In the previous sections, we have primarily been discussing the modeling related aspects of the framework and the software architecture corresponding to those. Here we briefly discuss a number of issues that generally relate to any simulation architecture. These issues include, timing and synchronization, event management, object and agent lifecycle management, message passing, visualization, simulation scenario-run management, distributed simulation management, and measurement and data logging.

All these issues significantly affect the fidelity and performance of the simulation and are very important aspects of any simulation tool. WEA has borrowed most of this infrastructure from an existing multi-agent simulation tool, i.e. the Re-configurable Flight Simulator (RFS) [8, 9]. WEA builds on this existing architecture, and its primary contribution is the architecture for comprehensively accommodating cognitive engineering models of the environment and the agents for multi-agent simulation. Except for timing and synchronization, all other utilities have been enhanced at different scales to accommodate for the environment and agent models proposed by WEA.

Timing and synchronization are key to the operation of this framework. The RFS and hence the WEA simulation uses three modes of timing, which includes uniform time step, asynchronous and asynchronous with resynchronization. The last mode is used for the highest fidelity simulations. In this mode, each agent informs the timing management unit about the next time when it would be best for the agent to updated. For example, while making a complex maneuver the aircraft agent may decide to user smaller time steps then when flying straight and level. The resynchronization utility maintains the fidelity relationship with all other agents that are affected by the presence of a particular agent in the system. This includes maintaining dependence lists for agents that should be updated before and after each update of any agent.

In the previous section, it was briefly stated that an agent could be following intelligent goals or just reacting to requests. Well, the timing module becomes critical here. That is, the simulation timing module calls a specific function in each agent, and when pursuing some intelligent goal, each intelligent agent will evaluate its context and knowledgebase on this call. And, if it wants to increase time-based fidelity of its behavior, it will reduce the update time span for the timing module.

To provide brief insights into other aspects of the simulation, there is an Object Data/Method Exchange utility at the simulation level that is used to set insight into each agents capabilities and its context. Distributed simulation management is done with the use of the High Level Architecture (HLA) [10]. Visualization is done through the provision of display modules that can either simply trace values of system variables, or provide complex displays such as the Air Traffic Control Display shown later in figure 5.

A measurement and analysis utility logs metric data in each design component and at the system performance level.

3.4 Summary of WEA Framework

WEA can computationally model and simulate large-scale and complex multi-agent socio-technical systems. WEA is a tool for engineering such systems through declarative manipulation of system design variables such as procedures, regulations and technologies in the work environment.

The framework provides a three-step process, which starts with modeling the work environment, generation of agent models, and simulation of usage scenarios for system analysis. Multiple aspects of the work environment such as its physical, functional and social aspects are represented as declarative and extensible structures. These declarative structures, together with environment aware agent models of proactive entities in the system, are employed in multi-agent simulations.

The design driven architecture of WEA enables system designers to directly tailor the design variables, without having to recode each agent for each system design change. System simulations of multiple agents interacting with each other and the models of the environment with embedded data collection for arbitrarily configured performance metrics enables detailed analysis of system design. Due to the structure and interaction preserving nature of these multi-agent simulations, the resulting designs can be easily applied to real world work environments and training of agents.

4 An Illustrative Example in Air Traffic Control

A systems procedure evaluation problem in air traffic control is being shown here to illustrate the WEA. When controlling the approach of aircraft to busy airports, the controllers may employ either Time Based Metering (TBM) or Miles In Trail Metering (MITM). TBM refers to procedures in which aircraft are instructed to reach a metering arc by a specific time. A metering arc is a hypothetical arc at a specific distance from a navigation fix, generally the fix that identifies the airport. MITM refers to procedures that line up aircraft in a trail behind each other as they approach the airport. In MITM aircraft follow a specific route, but there could be several of these routes approaching the airport and they may intersect each other. Only one of these procedures is in operation at one time.

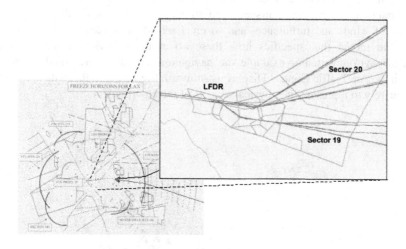

Fig. 4. The eastern approach sectors of the Los Angeles airport (LAX)

Controllers are continuously observing their displays and vectoring aircrafts to reduce congestion and to avoid any air hazards, while complying with these procedures. Pilots have to be able to follow instructions with regards to the procedures being used. When such a system is being evaluated for safety, the controller workload, the density of aircraft in airspace, air space violations and number and severity of vectoring commands for controlling aircraft are used to measure performance. Weather conditions, number and spatial configuration of aircraft in the airspace, their routes and their size are the independent variables that define the system operation scenarios.

To model and simulate such a system, human models of the pilots and controllers have to be created. These models should be able to comply with different procedures. They should possess different capabilities representing their training on specific procedures. The environment's physical aspects should be well represented. Dynamics of all equipment and technology should be modeled to significant fidelity. The procedures and regulations in the work-environment should also be modeled.

As can be easily deciphered, this problem calls for modeling multiple aspects of the work-environment and it requires modeling agents that are aware of this environment. Moreover, the environment design will change, and the agents should change correspondingly.

This domain problem was modeled and simulated using WEA. The simulation utilized real world scenarios from the eastern approach of the Los Angeles airport (LAX) (refer figure 4). The controllers in six control sectors were simulated (only three are shown in figure 4). Physical and Context-Process aspects of the work-environment were modeled. The following discusses these models.

4.1 Physical Environment

The physical environment in this domain consists of navigational data, weather models, terrain database and earth axis definitions. Navigational data consists of spatial data on airports, fixes, airspace definitions and navigational aids. Weather data consists of winds and turbulence, and so on. Each of these elements comes with a behavioral model that specifies how these affect the behavior of agents in this physical environment. For example the navigation aids are represented in specific ways in the Air Traffic Control Display as shown in Figure 5. All this data is specified declaratively to represent the physical aspects of the environment.

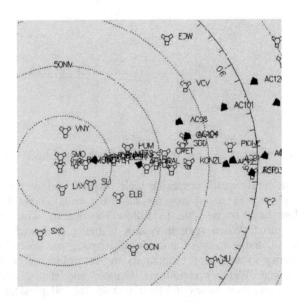

Fig. 5. The physical environment of the design problem. This figure shows the navigation aids and the simulated aircraft in the WEA simulation. This figure is a cut out of the Air Traffic Controller's display as simulated in WEA

4.2 Context-Process Structure

Since the domain problem involved evaluating procedures and regulations in different contexts, a context-process structure had to be devised as a new cognitive-engineering model for the work environment. Existing models were limited to task structures, but these could not used for this problem. The domain problem is not concerned with strategy, tactics formation or coordination on part of the agents [11], rather the problem is about designing procedures and regulations in the work environment applicable in specific contexts.

Thus a context-process structure had to be devised, each node of which specifies a process (procedure, action, or regulation) in the context in which it is applicable. The context is defined by the perceivable situation that should be fully understood for the application of the process. The objective of the agent at the time of applicability, i.e. the intention of the agent, is also needed to completely specify the context of the respective process.

Figure 6 illustrates a process node applicable for radar separation in the air traffic control domain. It specifies a restricted action or a regulation, where radar separation is prohibited for aircrafts in a specific range of air traffic controller's radar display (available in FAA regulations [12]). But this regulation is only applicable if the situational context is as defined in the node, i.e. positions of aircrafts in the display, the kinds of displays and the identification of each aircraft. The context is further made complete by the specification of the intention of the air traffic controller, i.e. radar separation. Note that this process node only restricts radar separation for the given situational context. The controller could apply visual separation for that situational context if there is no other process in the work environment restricting him/her from doing so.

Intention:		Lateral Radar Separation
World State:	Type of Radar Display	Primary Radar Targets, ASR-9/Full Digital Radar
	First Aircraft Radar Identification	Not Radar Identified
	First Aircraft Position	$\{x: x < 40\}$ x => distance from antenna
	Second Aircraft Radar Identification	Radar Identified
	Second Aircraft Position	$\{x: x < 40\}$ x => distance from antenna
Restricted Action:		Radar Separation in y, $\{y: y < 6\}$ y => distance of airspace in miles from the edge of display

Fig. 6. Air traffic control regulation for lateral radar separation showing restricting conditions. The intention and the world state together specify the context for this regulation

Each of these nodes is connected with other nodes through relations amongst intentions and the situational context of the processes. For example, a process applicable in a particular situation is still applicable in a situation, which may be better defined through additional variables or better defined range of values against which the situation is being evaluated. Similarly if the objective is to apply radar separation between aircraft, the processes applicable here are also applicable for vertical or lateral radar separation. Such relations between the process nodes construct the context-process environmental aspect to be used for the given design problem.

4.3 Models of Agents

As discussed in section §3.2, agents for this problem are generated based on generative models that specify parts of the environment visible to each agent. In this problem all physical elements are fully accessible to each agent through their technological equipment, i.e. the instruments in their aircraft. For example, an air traffic controller observes aircraft through radar displays and communicates with pilots only through specific channel frequencies. Therefore, if the radar cannot track an aircraft, the aircraft is not visible to the air traffic controller. Or if a pilot is not on a controller's frequency, the pilot and the controller cannot communicate.

The main issue here is the configuration of the environment for the applicable process, i.e. TBM or MITM, and the distribution of process nodes among the agents. As shown in figure 3, the pilot picks up capabilities from the environments description of the processes and the agent generative model that specifies what process the pilot should know. Thus the capability of flying a hold pattern is put into the pilot because one of the processes in the environment requires that the pilot should be able to fly a hold pattern when commanded to do so by the controller.

The hold pattern capability builds up on other capabilities such as flying straight and level and making standard rate turns. These capabilities are therefore also put into the Pilot agent. All agents, controllers and pilots, are created based on such environment aware agent generation models.

4.4 Simulation and Analysis

The simulations were run based on real-world scenarios from the LAX airport, based on the independent variables described in §4. These simulations are being validated for aircraft tracks and communication and maneuver events that actually took place in the real world.

5 Conclusion

This paper described the Work-Environment Analysis simulation framework. WEA computationally models large-scale and complex multi-agent socio-technical systems using multi-agent simulations. It is a framework for engineering such systems through design variables such as procedures, regulations and technologies in the work environment, as well as changes in the capabilities of their individual agents (usually human). The authors are currently demonstrating, through case studies, the utility of simulations and analysis done using WEA.

Even though this framework primarily contributes to the field of systems engineering by enabling detailed analysis of socio-technical systems, this paper demonstrates one major application of multi-agent based simulations.

To the field of multi-agent based simulation, the primary theoretical contribution is in the use of the *container* model for elements of the agents' environment and for the agents. This model can span multiple aspects of the work environment and enables dynamically generation of corresponding environment aware agents. This model also enables dynamically changing the scope of a simulation by employing a re-configurable set of models.

Amongst the major technological contributions is the provision of a versatile, scalable and extensible simulation toolkit that can be used to evaluate design changes to socio-technical systems. The framework is structure preserving to facilitate translation between the actual system being studied and agent-based simulation of its behavior, significantly decreasing the complexity of the design problem.

References

1. Vicente, K.M.: Cognitive work analysis: towards safe, productive and healthy computer based work. Lawrence Erlbaum Associates. (1999)
2. Simon, H.A.: Models of bounded rationality. The MIT Press, Cambridge, MA, (1982). Vol. 2
3. Davidsson, P.: Multi agent based simulation: beyond social simulation. In: Multi Agent Based Simulation. Springer Verlag LNCS series. (2000)
4. Hayes, C.C., Agents in a nutshell - a very brief introduction. IEEE Transactions on Knowledge and Data Engineering, (1999) 11(1).
5. McNeese, M.D., Vidulich, M.A. (eds.): Cognitive systems engineering in military aviation environments: avoiding cogminutia fragmentosa!. In: State of the Art Reports, Human Systems Information Analysis Center. (2002)
6. Annett, J., Stanton, N.A. (eds.): Task Analysis. First edition. Taylor and Francis. New York. (2001)
7. Johnson, H., Johnson, P.: Task knowledge structures: psychological basis and integration into system design. In: Acta Psychologica. Vol 78: 3 – 24. (1991)
8. Ippolito, C. A., Pritchett, A. R.: Sabo: A self-assembling architecture for complex system simulation. In: Proceedings of AIAA Aerospace Sciences Meeting and Exhibit, Reno, NV, (2000).
9. Pritchett, A. R., Lee, S. M., Corker, K. M., Abkin, M. A., Reynolds, T. R., Gosling, G., et al.: Examining air transportation safety issues through agent-based simulation incorporating human performance models. In: Proceedings of IEEE/AIAA 21st Digital Avionics Systems Conference, Irvine, CA, (2002).
10. High Level Architecture. Defense Modeling and Simulation Office, United State Department of Defense. https://www.dmso.mil/public/transition/hla/
11. Decker, K.S.: Environment centered analysis and design of coordination mechanisms. PhD Thesis. Department of Computer Science. University of Massachusetts, Amherst. (1994)
12. Air Traffic Control. Chapter 5, Section 5, Order 7110.65P. Federal Aviation Authority. http://www.faa.gov/atpubs/ATC/Chp5/atc0505.html#5-5-1

Layering Social Interaction Scenarios on Environmental Simulation

Daisuke Torii[1], Toru Ishida[12], Stéphane Bonneaud[3], and Alexis Drogoul[3]

[1] Department of Social Informatics, Kyoto University,
Kyoto, 606-8501, Japan
torii@kuis.kyoto-u.ac.jp, ishida@i.kyoto-u.ac.jp
[2] JST CREST Digital City Project,
[3] LIP6, Université Paris 6 8 rue du Capitaine Scott 75015, Paris, France
stephane.bonneaud@free.fr, Alexis.Drogoul@lip6.fr

Abstract. For an integrated simulation such as the natural environment affected by human society, it is indispensable to provide an integrated simulator that incorporates multiple computational models. We proposed a multi-layer socio-environmental simulation by layering the social interaction scenario on environmental simulation. For this simulation, we connect two different systems. One is a scenario description language Q, which is suitable for describing social interactions. Another is CORMAS, which models interactions between a natural environment and humans. The key idea is to realize a mapping between agents in different systems. This integration becomes possible by the salient feature of Q: users can write scenarios for controlling legacy agents in other systems. Moreover, we find that controlling the flow of information between the two systems can create various types of simulations. We also confirm the capability of CORMAS/Q, in the well-known Fire-Fighter domain.

1 Introduction

Global warming, acid rain, forest fire, earthquake disaster, etc. are the big topics of environment and disaster in the world, and information technology is expected to solve these problems. For example, an integrated simulation of human societies and natural environments, named Socio-Environmental Simulation, is able to help with policy and training for these problems. Some systems for MABS do not give specific computational models of agents for general uses. For example, CORMAS [3] is one of the well-known systems for socio-environmental simulation. For realizing social interactions of agents, users have to construct models from scratch.

There are two streams for coping with this problem. One is the connection of computational models, which is called "Docking" [1]. This aims to compare two computational models and confirm its adequacy, and to explore the possibility of combined models. Another is construction of multi-layer architectures. A simulation for some purpose is flexibly realized by connecting two different systems which have different features. For example, the RoboCup simulator gives protocols for

P. Davidsson et al. (Eds.): MABS 2004, LNAI 3415, pp. 78–88, 2005.

connecting various agent models to the soccer server [9]. Q is a language for designing social interactions of agents by attaching Q to already existing agent systems. This language provides a description for non-computer professionals and has generality for connecting every agent system in any domain. This has been confirmed by connection with FreeWalk and Microsoft Agents [5, 7, 8].

In this paper, we propose a multi-layer socio-environmental simulation. As a comprehensive example, we connect CORMAS, intended for describing interactions between natural environment and humans, and Q, developed for describing social interactions. CORMAS is widely used to simulate natural resource management [2, 10]. A natural environment is simulated using cellular automata and observations/actions of humans are defined as agents. Q is a language to express extended finite state automata, which are often used for describing protocols among agents or interaction scenarios. Q has been applied to several social psychological simulations [7, 8].

This connection is not only an integration of two computational models, but also a construction of a multi-layer architecture by integrating two different legacy systems. In this case, not only cellular automaton is connected with extended finite state automaton, but also IPC (see 2.1) is available and participatory simulation is easily realized by Q on CORMAS. This connection is made possible by the salient feature of Q: users can write scenarios for controlling legacy agents in other systems. Moreover, we propose a method that various types of simulation can be generated by controlling the flow of information in the channel connecting the two systems. To connect Q and CORMAS, we had to address the following two issues.

Functional Distribution
It is necessary to define a mapping of corresponding agents in both systems. In the case of CORMAS/Q, observations and actions in the natural environments are executed within the CORMAS module while the interaction scenarios are performed in the Q module. Mapping is made possible by the salient feature of Q: users can write scenarios for controlling legacy agents in other systems (like CORMAS).

Time Management
Ensuring time consistency in both modules is a challenge. In the case of CORMAS/Q, to ensure time consistency, we took the approach that only CORMAS manages time. Q has no responsibility for time management. To make this possible, we assume that Q interprets scenarios far more rapidly than CORMAS simulates observations/actions of agents (see Section 4.1 for details)

The following sections explain CORMAS/Q, and show how the well-known Fire-Fighter example (Cohen [4]) validates a socio-environmental simulation of CORMAS/Q.

2 Background

2.1 Q^1

Q is suitable for describing complex social interactions and has been applied to social psychological simulations in a virtual space [8] such as evacuation simulations. Q can

---

[1] Q is available from http://www.digitalcity.jst.go.jp/Q/index.html

describe interaction scenarios between (legacy) agents (See Table 1). This feature makes it possible to control the scenario execution of a large number of agents by attaching Q to already existing agent systems. The computational model behind a Q scenario is an extended finite state automaton, which is commonly used for describing communication protocols. By using Q, users can directly create scenario descriptions from extended finite state automata.

In Q scenarios, we can use sensing functions (cues) and acting functions (actions) provided by already existing agent systems. Scenarios are interpreted by Q, while cues and actions are executed by the legacy agent systems. The mapping between cues/actions in the Q scenarios and those in agent systems is given by the Q Agent Adapter. Moreover, Q provides an end user language called *Interaction Pattern Card (IPC)*. Since IPC is domain dependent and implemented using Excel, it enables non-computer professionals to create scenarios easily. Scenarios described in IPC are translated into Q by the IPC translator.

It is possible to replace Q controlled software agents by user controlled avatars, i.e. participatory simulation is easily realized by replacing some of the agents by humans. Simulations are carried out in an event driven fashion. The Q interpreter has no responsibility for time management: time is controlled by the agent systems that execute the bodies of cues and actions.

Table 1. Q and CORMAS

	Q	CORMAS
Goal	• A language for describing interaction scenarios among (legacy) agents • A large number of agents can be controlled by attaching Q to already existing agent systems.	• System for simulation whose models of coordination modes between individuals and groups who jointly exploit the resources • The changes of natural environment and the observations/actions of agents are defined
Computational Model	• Extended finite state automaton	• Cellular automaton
Description	• Scenarios are described, using sensing functions (cues) and acting functions (actions). • *Interaction Pattern Card (IPC)* enables non-computer professionals to create scenarios. • Participatory simulation is easily realized by replacing a part of software agents by human controlled avatars.	• At every step (unit time), the simulation is executed by describing functions of each cell and agent. • Modeling tools (space, agent, communication between agents), a management tool (simulation) and visualization tools (communication between agents, statistical information) are available. • It is optionally possible to import the spatial map from GIS (Geographic Information Systems) like MapInfo, ArcView.
System	• Event driven. • The Q interpreter has no responsibility on time management.	• Time driven. • At every step (a unit time), the diffusion between neighboring agents is calculated. • Time management is done by allocating appropriate time length to each step.
Application	• Social psychological simulation. • Evacuation simulation.	• Resource management of water, wood and pasture. • Multiple uses of land and resources.
Implementation Language	• Scheme	• Smalltalk

Q is an extension of Scheme, a Lisp programming language dialect. Concurrent execution of scenarios of multiple agents is realized easily by using Scheme's *continuation* for controlling process switching.

2.2 CORMAS[2]

The goal of CORMAS (COmmon pool Resources and Multi-Agents Systems) is to build ˋsimulation models of coordination modes between individuals and groups who jointly exploit the (same) resources [3]. In CORMAS, users can define the diffusion of environmental changes, and agents' observations/actions with regard to the environment (see Table 1). The computational model behind CORMAS simulations is a cellular automaton. A natural environment is modeled as a two dimensional mesh, and the diffusion between neighboring cells is calculated at each unit time (called a *step*). Modeling tools for space, agent and communication, a simulation management tool, and visualization tools are available. It is possible to import map information from GIS (Geographical Information Systems) like MapInfo[3] and ArcView[4]. CORMAS has been applied to renewable resource management of water, wood and pasture [10], natural resources marketing systems, and multiple uses of land and resources [2].

The CORMAS system is time-driven. At every step, the diffusion from neighboring cells is calculated, and agents' observations/actions are performed. Time management is done by allocating the appropriate step duration. CORMAS provides two types of simulations. In synchronous simulations, changes of all entities (cells, agents) are committed at the end of each step. In asynchronous simulations, on the other hand, the changes are immediately reflected in ongoing calculations. Users can construct a model of agents and natural environments by using prepared Smalltalk classes. Users can add new classes as necessary.

3 Multi-layer Socio-environmental Simulation

3.1 Architecture

The basic idea of multi-layer socio-environmental simulation is combining social simulation with environmental simulation. This simulator consists of two layers: social and environmental simulation layers. Social simulation offers decision making, negotiation, and collaboration. Environmental simulation realizes the diffusion of environmental changes and agents' reactions.

Social interaction and environmental diffusions are executed as follows. A scenario of social interaction (ex. decision making process) is executed in the *social simulation layer*, but communication acts, like messaging, are performed by agents in the *environmental simulation layer*. This approach is needed since all real world actions must be performed in one layer to solve the time synchronization problem. If the

---

[2] CORMAS is available from http://cormas.cirad.fr/indexeng.htm
[3] http://www.mapinfo.com/
[4] http://www.esri.com/software/arcview/

social simulation runs much faster than the environmental simulation, it is possible for the environmental simulator to manage time.

The environmental diffusions are calculated in the environmental simulation layer, while agents' observations/actions to natural environments are controlled in the social simulation layer. If agents execute actions in natural environments, the diffusion of the influence of the actions is also calculated in the environmental layer. The agents in the social simulation layer then acquire the results of the environmental diffusion.

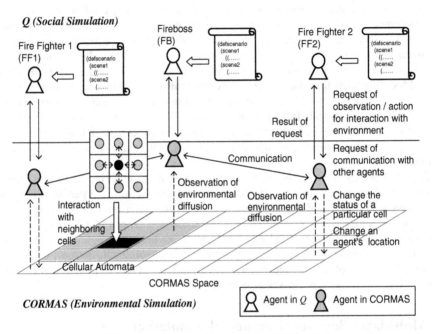

Fig. 1. Fire Fighting Simulation by CORMAS/Q

3.2 Connection Control

Various types of simulation can be realized by controlling the flow of information through the channel connecting the different layers; the legacy systems do not need to be altered. For example, there are three types of information flowing through the connection channel as follows:

A)   observation of environment
B)   action to environment
C)   communication with other agents

The following two operations can be performed on the above information.

I)   to lose information (accuracy)
II)   to delay information transmission (delay)

These operations are defined by the simulation user in the connection description as follows:

Communication Channel

The accuracy and timeliness of information flow in organization needs to be defined if the simulation is to be realistic. In the example of the fire fighter model (see Fig.1.), several fire fighters work under the command of the fireboss. It is possible to control the flow of information between the fireboss and the fire fighter and between fire fighters.

For example, when the accuracy of the flow of information from the fireboss to a fire fighter is set to 0.8 and the delay is set to 3, one in five messages from the fireboss to the fire fighter is lost and all messages are delayed for 3 unit times. For another example, the accuracy with which the fire fighter acquires environmental information could be set to 0.5 to reflect the difficulty of gathering such information.

Efficiency of Information Gathering

The organizational ability of information gathering is defined. In the fire fighter model, the fireboss might have much lower accuracy and timeliness than the fire fighters.

Efficiency of Environmental Actions

Organizational efficiency of environmental actions is also defined in terms of accuracy and delay. In the fire fighter model, the efficiency of fire fighting (the action) in the environment might be only 70%.

What we want to stress here is that the connection between the two modules (social simulation and environmental simulation) can be realized without modifying the connected simulators. All that is needed is to create a connection description and applying a few simple operations.

4   Connecting Q and CORMAS

4.1 Q and CORMAS

To put the architecture shown in the previous section into practical use, we actually connected Q to CORMAS and applied the resulting ensemble.

Q scenarios are used to describe sensing (Cue) and acting (Action) acts in the environment. Simultaneously, the receiving (Cue) and sending (Action) of messages between agents are described. The agents in Q are used only for the interpretation of Q scenarios; the corresponding agents in CORMAS actually execute the body of Cue and Action. Therefore, the body of Cue and Action should be defined in the CORMAS module. When the execution of Cue and Action is required, the request should be transferred from Q to CORMAS and the execution result should be transferred from CORMAS to Q.

What should be considered in this connection is time management. The obvious approach is to manage time independently in Q and CORMAS and require the Connection Control Module to establish time consistency. However, since Q scenarios are carried out in an event driven fashion while CORMAS simulations are carried out in steps, control would be needlessly complicated and CORMAS/Q would have to be modified extensively. Our approach is for CORMAS to manage time completely, and all observations/acts that take actual time are requested from Q to CORMAS. This means that Q is oblivious of the time management performed by

CORMAS under the assumption that CORMAS operations proceed smoothly because Q operations are far faster than those in CORMAS

In CORMAS, spatial diffusion and observations/actions are calculated at every step. For example, when an action requested by a Q scenario is executed, the spatial diffusion is calculated at the following step in CORMAS. After this, the observation requested by the Q scenario triggers feedback from CORMAS to Q.

4.2 Connection Control Module

For connecting Q and CORMAS, we implemented the Connection Control Module which consists of the Q Connector/CORMAS Connector to extend the Q module/CORMAS module. Q Connector has two roles. One is to write execution requests of Cue/Action to CORMAS side. The other is to receive the execution results and return them to the Q interpreter. The CORMAS Connector also has two roles. One is to receive execution requests from the Q module and call the corresponding CORMAS method. The other is to return the execution results to Q side.

5   Fire Fighter Domain

5.1 Problem

To validate our approach, we ran a simulation using the model of a forest fire in Yellowstone National Park developed by the Phoenix Project at the University of Massachusetts Amherst in 1989 [4].

In this problem, two types of agents exist. One is a fireboss who gives a direction based on the overall environmental information. The other is a fire fighter who follows the fireboss's direction and moves in the environment. The fire fighter is a bulldozer and builds a line around continuously burning fires to prevent the fires from spreading.

In the environment, each cell has a type such as a river, a plain, a road, wood, and fire. Fires spread in irregular shapes and at variable rates, determined by ground cover, elevation, moisture content, wind speed and direction, and natural boundaries.

This simulation starts when the fireboss is alerted to a new fire and dispatches two fire fighters to a rendezvous point which is determined by a calculation based on to the location of the fire and wind speed/direction. Next, the fireboss calculates a route for the fire fighters and instructs them to follow the given path and encircle the fire. They are given clockwise and counterclockwise encircling directions. When they find a fire on their route, they can change their route by themselves. If the fire threatens them due to an abrupt environmental change, they move away from the fire. When such a route change is reported to the fireboss, he sends back a better route plan calculated from the location of the fire fighter and overall environmental information. The fire fighter changes his own route plan as suggested by the fireboss when the difference between his own plan and the fireboss's plan is large. When a fire fighter encounters a natural boundary like a river or a fire line created by the other fire fighter, he asks the fireboss for the next action. If a fire leaps outside the fire circle, the fireboss gives the fire fighters a new route plan. When the fire is enclosed and flying sparks are insignificant, the fireboss concludes that the mission is finished and stands down the fire fighters. In this way, the fire is controlled by the interaction of the fireboss and the fire fighters.

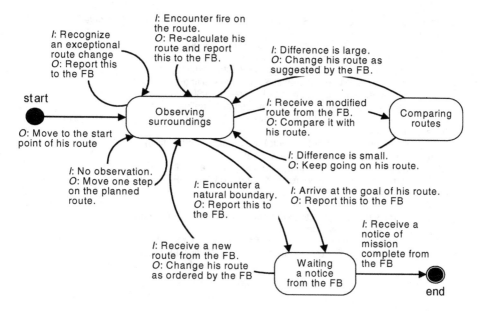

Fig. 2. State Transition Diagram of Fire Fighter

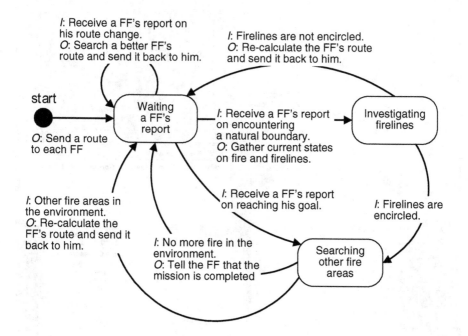

Fig. 3. State Transition Diagram of Fireboss

```
(defscenario fireboss  (&pattern ($FF_name "") ($FF_info "") ($env_info "") ($route ""))
    (Waiting-FF-report
        ((?route_change_report :from $FF_name :info $FF_info)
         (!get_env_info :info $env_info)
         (!calculate_route :info $FF_info :info $env_info :result $route)
         (!send_route :to $FF_name :result $route) (go Waiting-FF-report))
        ((?natural_boundary_report :from $FF_name :info $FF_info)
         (!get_env_info :info $env_info) (go Investigating-firelines))
        ((?goal_arrival_report :from $FF_name :info $FF_info) (go Searching-other-fire)))
    (Investigating-firelines
        ((?firelines_encircled :info $env_info) (go Searching-other-fire))
        ((?firelines_not_encircled :info $env_info)
         (!calculate_route :info $FF_info :info $env_info :result $route)
         (!send_route :to $FF_name :result $route) (go Waiting-FF-report)))
    (Searching-other-fire
        ((?no_more_fire :info $env_info)
         (!send_mission-complete :to $FF_name) (go Watching-FF-report))
        ((?other_fire :info $env_info)
         (!calculate_route :info $FF_info :info $env_info :result $route)
         (!send_route :to $FF_name :result $route) (go Waiting-FF-report))))
```

Fig. 4. *Q* Scenario for Fireboss

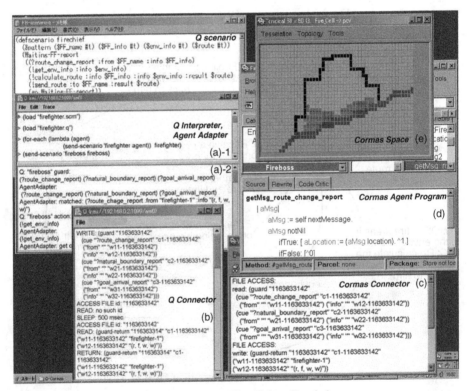

Fig. 5. Screenshot of Fire Fighter Simulation

5.2 Scenario

Fig. 1 illustrates the implementation of this problem in CORMAS/Q. In CORMAS, we implemented a simulation in which fires spread minute by minute, using a model that contains ground information like wood, a fire line and a river and weather information of wind speed and direction. On the other hand, the agents of the fireboss and the fire fighters are placed in both Q and CORMAS; Q controls agent execution in CORMAS. There are two types of interaction in the Q scenario. One is social interaction between the agents. The other is environmental interaction between the fire fighters and the environment.

The behavioral plans of the fire fighter and fireboss, already mentioned, are expressed in a state transition diagram (see Fig. 2, Fig. 3). Fig. 4 shows the Q scenario of the fireboss. The Q scenario was directly coded from the state transition diagram. It is easy to understand the correspondence between the problem explanation of Section 5.1 and the state transition diagram of Fig. 2 and Fig. 3, and the scenario description of Fig. 4.

5.3 Result

Fig. 5 shows a screenshot of this simulation. In the simulation, a small fire is started in the middle of the space and the fire is driven by a north wind. The initial location of the fire fighter is the north of the fire. We could have the result that the model intends with various strength of the north wind. The fire slowly spreads with weak wind (each cell changes its state to fire with less than 30% when its north cell is fire and with less than 15% when its north-east and north-west are fire. The possibility that a cell catches fire from the north fire was set to double of the north-east and the north-west fire.), which brings a result that the fire is encircled by only the fire line. To the contrary, the fire quickly spreads with strong wind (in many cases, the probability is more than the one of the weak wind), which brings a result that the fire is encircled by the fire line and the natural boundary (river). The CORMAS map (Fig. 5 (e)) shows the latter result. Fig. 5 (a) and (b) show the execution logs of Q, and Fig. 5 (c) and (d) show the execution logs of CORMAS.

In any strength of the wind, the fire was contained as desired by the scenario. It is necessary to try the other types of simulation except this fire fighter model (as noticed in the conclusion), but we consider that this result verifies the design of the scenario and the capability of multi-layer socio-environmental simulation.

6   Conclusion

In this paper, we proposed a multi-layer socio-environmental simulation by layering social interaction scenario on environmental simulation. As a concrete example of this idea, we combined two multiagent systems. One is Q, whose computational model is an extended finite state automaton for modeling of agents' social interaction. The other is CORMAS, whose computational model is a cellular automaton for modeling of an environment.

In the integrated simulation system, all observations/actions that take time to complete are executed in CORMAS. As a result, time management is performed only in CORMAS. This integration has been made possible by the feature of Q: users can write social interactions to control legacy agents. Moreover, various types of simulation can be generated by controlling flow of information in the channel connecting the two systems. By combining Q and CORMAS, we can integrate social and environmental simulations. We implemented a well-known fire fighter example in CORMAS/Q, and confirmed its ability to handle fairly complex simulations.

We are currently planning two socio-environmental simulations with CORMAS/Q. One is a joint project with Disaster Prevention Research Institute of Kyoto University to apply CORMAS/Q to participatory simulations of rescue activity under a flood situation. Since it is possible to replace Q controlled agents by user controlled avatars, CORMAS/Q has the ability to support participatory simulations (simulations involving collaboration between software agents and humans). Another is a joint project with IRRI (International Rice Research Institute) Thailand Office and CIRAD to apply CORMAS/Q to participatory simulations of agricultural economics. Furthermore, since Q scenarios are based on extended finite state automata, machine learning can, by applying the data from participatory simulations, improve the scenarios. The results will be reported in the near future.

References

1. Axtell, R. L, Epstein, J. M. and Cohen, M. D.: Aligning Simulation Models: A Case Study and Results. *Computational and Mathematical Organization Theory*, vol. 1, 123-141, 1996.
2. Bah, A., Touré, I. and Le Page, Ch.: An Agent-Based Model tool for multi-agent simulation for Understanding the Multiple Uses of space Land and Resources around a Drilling Sites in the Sahel. *In Proceedings of Modsim 2003 International Congress on Modeling and Simulation*, 1060-1065, 2003.
3. Bousquet, F., Bakam, I., Proton, H. and Le Page, C.: Cormas: common-pool resources and multi-agent Systems. *Lecture Notes in Artificial Intelligence* 1416, 826-838, 1998
4. Cohen, P. R., Greenberg M. L., Hart, D. M. and Howe, A. E.: Trial by fire: Understanding the design requirements for agents in complex environments. *AI Magazine*, Vol. 10 No. 3, 32-48, Fall 1989.
5. Ishida, T.: Digital City Kyoto: Social Information Infrastructure for Everyday Life. *Communications of the ACM (CACM)*, Vol. 45, No. 7, 76-81, 2002.
6. Ishida, T.: *Q*: A Scenario Description Language for Interactive Agents. *IEEE Computer*, Vol.35, No. 11, 54-59, 2002.
7. Murakami, Y., Ishida, T., Kawasoe, T. and Hishiyama, R.: Scenario Description for Multi-Agent Simulation. *In Proceedings of International Joint Conference on Autonomous Agents and Multiagent Systems (AAMAS-03)*, 369-376, 2003.
8. Nakanishi, H., Nakazawa, S., Ishida, T., Takanashi, K. and Isbister, K.: Can Software Agents Influence Human Relations? Balance Theory in Agent-mediated Communities. *In Proceedings of International Joint Conference on Autonomous Agents and Multiagent Systems (AAMAS-03)*, 717-724, 2003.
9. Noda, I. and Stone, P.: RoboCup Soccer Server and CMUnited: Implemented Infrastructure for MAS Research, *Infrastructure for Agents, Multi-Agent Systems, and Scalable Multi-Agent Systems*, Wagner, T., Rana, O., ed., 94-101, 2001.
10. Perez, P. and Ardlie, N., Kuneepong, P., Dietrich, C., Merritt, W. S.: CATCHCROP: modeling crop yield and water demand for integrated catchment assessment in Northern Thailand. *Environmental Modelling and Software*, Vol. 17, No. 3, 251-259, 2002.

Change Your Tags Fast! – A Necessary Condition for Cooperation?[1]

David Hales

Department of Computer Science, University of Bologna, Italy
dave@davidhales.com

Abstract. Several tag models with intriguing properties have been advanced recently. But currently there is little detailed understanding of the underlying processes. Specifically it is not know what (if any) are the necessary conditions for tag systems to produce high levels of cooperation. We identify, for the first time, what appears to be a necessary condition that previous tag models *implicitly* contained. It appears that, in general, for tag-based systems to support high levels of cooperation tags must mutate faster than strategies because cooperative tag groups need to spread (by mutation of tags) before free riders (by mutation on strategies) invade the group. We test this theory with simulation.

1 Introduction

Tags are markings or social cues that are attached to individuals (agents) and are observable by others [11]. They evolve like any other trait in a given evolutionary model. The key point is that the tags have no direct behavioural implication for the agents that carry them. Through indirect effects, however, they can evolve from initially random values into complex ever changing patterns that serve to structure interactions between individuals.

In the computational models discussed here tags are modelled using some number (either a binary bit string, a real number or an integer). When agents interact they preferentially interact with agents possessing the same (or similar) tag value. One way to visualize this is to consider a population of agents partitioned between different colours. Each agent carries a single colour. In a system with only 3 different possible tag values we could think of this as each agent carrying a flag of red, green or blue. Agents then preferentially interact with agents carrying the same colour (forming "interaction groups"). When agents evolve (using some form of evolutionary algorithm) they may mutate their tag (colour). This equates to moving between interaction groups.

In the models presented here, tags take on many possible unique values (by say using a real number, there are many possible unique tags rather than just 3 colours)

---

[1] This work partially supported by the EU within the 6th Framework Programme under contract 001907 (DELIS).

P. Davidsson et al. (Eds.): MABS 2004, LNAI 3415, pp. 89–98, 2005.

however, the basic process is the same – agents with the same tags preferentially interact and tags evolve like any other genotypic trait.

Another way to think of tags is that some portion of the genotype of an agent is visible directly in the phenotype but the other agents.

Hales [3] advanced a model, using binary tag strings that demonstrated the evolution of cooperative interactions in the *single round* Prisoners Dilemma (PD). Further work [17] showed the emergence of altruistic giving behaviour and the evolution of cooperation and specialization [5][2].

These latter models are important because they advance a novel mechanism for evolving coordinated and cooperative interactions between unrelated agents that have *no knowledge of each other and have never met previously*. This obviates the need for repeated interactions [20], "genetic" relatedness [10], "image scoring" [15] or strict spatial relationships [14] in the production of cooperation. Tag mechanisms therefore have potential engineering applications where these other methods are not applicable (see below).

Although the general mechanism by which tags produce these results appears to be the result of a dynamic group formation and dissolution process [3, 16, 19] with selection appearing to occur at the group-level, there has been little analytical or empirical exploration of this hypothesis.

2   Some Previous Tag Models

There have been a number of tag models implemented previously. All generally show how higher-than-expected levels of cooperation and altruism are produced when tags are employed. In all cases the models implement evolutionary systems with assumptions along the lines of the replicator dynamics (i.e. reproduction into the next generation proportional to utility in the current generation, no "genetic-style" cross-over operations but low probability mutations on tags and strategies during reproduction).

Riolo [16] gave results of expansive and detailed studies applying tags in a scenario where agents played dyadic (pair wise) Iterated Prisoner's Dilemma games (IPD). Tags (represented as a single real number) allowed agents to bias their partner selection to those with similar tags (probabilistically). He found that even small biases stimulated high levels of cooperation when there were enough iterations of the game with each pairing.

In Hales [3] a tag model was applied to a *single round* PD. Again interaction was dyadic. Tags were represented as binary strings. Pairing was strongly biased by tag identity (rather than probabilistic similarity). In this model very high levels of cooperation were produced between strangers in the one shot game.

In Riolo et al [17] a tag model was applied to a resource-sharing scenario in which altruistic giving was shown to emerge. Agents were randomly paired (some number

---

[2] It should be noted that the conclusions of these further studies have been questioned [18, 2]. Essentially the scenarios do not bear too close a comparison to a PD because there is no dilemma.

of times) and decided if to give resources or not. The decision to give was based on tag similarity mediated by a "tolerance gene" as well as the "tag gene" (both represented as real numbers). The utility to the receiving agent of any given resource was greater than to that of the giving agent. It was shown that if each agent was paired enough times in each generation and the cost / benefit ratio was low enough then high levels of cooperation were found.

In Hales and Edmonds [6, 7] tags were applied to a simulated robot coordination scenario, originally given by Kalenka and Jennings [13], producing high levels of cooperative help giving.

2.1 Mutation in the Models

We will now describe in, a little detail, how mutation was applied to the agents in each of the above models. We will not discuss the specific details of the reproduction process since we do not consider this relevant to the focus of this paper (variants of "roulette wheel" selection and "tournament selection" were used, and these produced probabilistic selection into the next generation following the replicator dynamics assumptions stated earlier). Neither will we focus on the interactions or specific payoffs applied in each model, suffice to say all models capture some kind of collective coordination / cooperation problem in which cheating or free riding is possible.

In order to examine and compare mutation schemes we make a distinction between the mutation rate applied to the tag and that applied to the strategy. In all cases agents are represented in the models using sets of artificial "genes" (some set of data types) that are mutated when copied into the next generation.

The descriptions of the models all explicitly state that the mutation rate applied to the tag and the strategy is *the same* (some probability). We label this rate m. However, models vary in the *mutation operation* applied with probability m and in the way they represent tags and strategies. It is this variation of mutation operation and tag / strategy representation that can hide what is best understood as a variation in mutation rate.

2.1.1 Bit String Representation of Tags with Simple Strategies

In Hales [3] tags are represented as fixed length bit strings and strategies as a single bit (either to cooperation in the single-round PD or to defect). The mutation rate is m = 0.001 and the population size is p = 100. Since each agent is completely represented by a binary string the mutation operation is simply to flip each bit with probability m (both tag and strategy bits). It would superficially appear that strategy and tag are therefore mutated at the same rate and in the same way. However the results of the paper show that high cooperation only occurred when the number of tag bits L was large (L = 32 or more). In these cases the tag is more prone to mutation than the strategy because it contains more bits. Any change in the tag effectively creates a new distinct tag because pairing in the model is based on tag identity not similarity. So the effective mutation rate on the tag as a whole is $1-(1-m)^L \approx 0.0315$ (more than 30 times that on the strategy).

2.1.2 Real Number Representation of Tags with Simple Strategies

In Riolo et al [17] each agent is composed of two real numbers - one representing its tag and one representing a so-called "tolerance". The tolerance is a kind of "proxy strategy". Essentially (simplifying) a smaller tolerance value means a less cooperative agent. Mutation is applied to bother the tag and tolerance with probability m = 0.1. Again it appears that both are being mutated with the same rate. However, the mutation operation applied to the tag is to replace it with a random value drawn informally from the range but the tolerance has Gausian noise (of mean 0 and standard deviation 0.01) added to it. So tags, when mutated, get new values chosen randomly from the range but tolerances get modified by small values. Simplifying the analysis somewhat, we could expect the absolute average tag change amount to be $\approx$ 0.333 when mutation is applied. Since m = 0.1 we might characterize the average overall tag change amount to be $\approx$ 0.0333. In the case of tolerance we can see that the absolute average change would be almost two orders of magnitude lower ($\approx$ 0.0008).

2.1.3 More Complex Strategies

In both Riolo [16] and Hales and Edmonds [6] our analysis becomes slightly less straightforward. In both cases strategies are composed of multiple "genes" which do not relate to simple strategies of unconditional cooperation or selfish behaviour. This is in part due to the scenarios. In [16] agents play the IPD with agents having similar tags for a number of rounds. The level of cooperation produced is not high and constant but fluctuates into periods of high and low cooperation. Tags are represented by single real values [0..1], strategies by triples of real values <i, p, q> each a probability capturing a probabilistic IPD strategy space (i is the probability of cooperation for the first round, p the probability of cooperation if in the previous round the other agent cooperated, q the probability of cooperation if the other agent defected on the previous round). So a space comprising tit-for-tat as well as pure defection and pure cooperation is formed (along with probabilistic variants). The mutation rate m = 0.1 is the same for each trait as is the operation (adding Gaussian noise with mean 0 and standard deviation 0.5). Here we have an interesting counter-point to the previous model [3] where we stated (above) that because the tag was split in several parts the effective mutation rate was higher than the strategy. Here, we have the reverse, so surely this suggests that the mutation rate applied to the tag is lower than that applied to the strategy? In one sense this is true. However, what is important is not the representation as such, the stored value, but how that value *relates to behaviour*. Since the strategy is a triple, in which pure cooperation is represented as all values being 1 and pure defection all values being 0, the relationship between mutation and the resultant change in strategy is not simple. However we can note that the probability of going from a triple of zeros to a triple of ones (from pure defection to pure cooperation) in a single mutation event is approaching zero. However, since we are talking about IPD not just a single round interaction the situation is more complex and we leave detailed treatment to a future paper[3].

---

[3] The cooperation found here [16] was not for the single interaction kind given in [3] and [17]. Indeed one of the findings of the paper was that the given model did not produce cooperation in the single round game.

In Hales and Edmonds [6] simulated robots work in teams to unload trucks in a warehouse. Here again we have a strategy composed of multiple parts. In the model tags are represented as single cardinal values [1..500] and strategies as pairs of binary values. Again the way the strategy effects behaviour is complex and moderated by the scenario. However, to simplify, a strategy represented by bit values "11" represents full cooperation whereas a value of "00" represents completely selfish behaviour. Mutation is applied to the triple of traits with rate m = 0.1. The mutation operation is to replace the existing value with another value chosen uniformly randomly over the space. Again simplifying things a little we can say that the probability of a strategy changing from 11 to 00 (or vice versa) is the probability that two bits are replaced with their compliment $0.25(m^2) = 0.0025$. The probability of a completely new tag (again tags are distinct, matching on identity) is $0.998(m) = 0.0998$.

2.1.4 Summary

So in all these cases it appears tags change more quickly than strategies under an algorithm that presents a uniform mutation rate. Of importance (as stated before) is the representation of tags and strategies and mutation operators taken together with the mutation rate. Only by considering all these factors can an underlying average relative rate of change be estimated between the two entities (tag and strategy). In each case when we do this we find that the tag changes much more quickly than the strategy. Next we advance a hypothesis based on this.

3   Hypothesis and Theory

From our analysis of the mutation schemes in the previous tag models we now advance a qualitative hypothesis concerning a necessary condition for tag models to produce high cooperation in one-time interactions: In general for tag based systems to support high levels of cooperation tags must mutate faster than strategies. We can also state a qualitative "mini-theory" to explain this: Cooperative tag groups need to spread (by mutation of tags) before free riders (by mutation on strategies) invade the group[4].

We don't have a quantitative complement to these two statements. It would appear that in order to determine the specific numbers in a specific scenario (model) we would need to consider the nature of the tag space, the nature of the strategy space and the way agents specifically interacted (the game). This is an aspect of on-going work.

3.1   Testing the Hypothesis

In order to test our hypothesis we implemented a new (minimal) tag model in which agents play single rounds of PD. We consider the result of high cooperation in the single round PD to be the most significant result so far advance for tags. Additionally the scenario is well understood and there are many existing models that allow for

---

[4] For an illustration of the tag-group process. We refer interested readers to [3].

comparison. The singe-round PD captures, in a minimal way, many of the essential features of the problems of cooperation in collective interactions. In our new model we varied the relative mutation rate between the tag and strategy to examine if this had an effect on the amount of cooperation produced. The model and results are described below but firstly we briefly outline the single-round PD.

3.2 The Prisoner's Dilemma

The Prisoner's Dilemma (PD) game captures a scenario in which there is a contradiction between collective and self-interest. Two players interact by selecting one of two choices: Either to "cooperate" (C) or "defect" (D). For the four possible outcomes of the game players receive specified payoffs. Both players receive a reward payoff (R) and a punishment payoff (P) for mutual cooperation and mutual defection respectively. However, when individuals select different moves, differential payoffs of temptation (T) and sucker (S) are awarded to the defector and the cooperator respectively. Assuming that neither player can know in advance which move the other will make and wishes the maximize her own payoff, the dilemma is evident in the ranking of payoffs: $T > R > P > S$ and the constraint that $2R > T + S$. Although both players would prefer T, only one can attain it. No player wants S. No matter what the other player does, by selecting a D move a player ensures she gets either a better or equal payoff to her partner. In this sense a D move can't be bettered since playing D ensures that the defector cannot be suckered. This is the so-called "Nash" equilibrium for the single round game. It is also an evolutionary stable strategy for a population of randomly paired individuals playing the game where reproduction fitness is based on payoff. So the dilemma is that if both individuals selected a cooperative move they would both be better off but both evolutionary pressure and game theoretical "rationality" selected defection.

3.3 The TagWorld Model

Our TagWorld model is a variation on [3]. We use a single real number to represent tags rather than a binary string. What is new is that we explicitly vary the mutation rate applied to the tag while keeping the rate constant for the strategy. Agents are represented by a single binary (the strategy bit) and a single real number in the range [0..1] (the tag). The strategy bit represents a pure strategy: either unconditional cooperation or unconditional defection. Initially the population have their strategy and tag values set to randomly with uniform probability over the space of all possible values. The following evolutionary algorithm is then applied.

In each generation each agent (a) is selected from the population in turn. A game partner is then selected. Partner selection entails the random selection of another agent (b) from the population such that (a) $\neq$ (b) but the tags of (a) and (b) are identical. If no agent exists with identical tags to (a) then (b) is selected at random from the entire population regardless of tag value. Consequently (a) will always find a partner even if its tag does not match any other agent in the population. During game interaction (a) and (b) invoke their strategies and receive the appropriate payoff. After all agents have been selected in turn and played a game a new population is asexually reproduced. Reproductive success is proportional to average payoff. The entire population of agents is replaced using a "roulette wheel" selection method [1][5].

3.3.1 Parameters Used in the Model

For the results presented here we used similar parameters to [3], though here we did not execute a scan over the parameter space. The population size was N = 100 and the number of generations for each run of the model was 1000. The PD payoffs were T = 1.9, R = 1, P = S = 0.0001. These values were selected to give a very high incentive to cheat (T is high and P and S are low). P and S were selected as a small value but greater than zero (indicating a very small chance for agents, with Sucker or Punishment payoffs, of reproduction). If a small value is added to P (enforcing T > R > P > S) results are not significantly changed.

For the strategy bit the mutation rate was fixed constant at $m = 0.001$ (a low value). But for the tag a mutation factor f was applied to m changing the mutation rate. We varied f from [0..10] in increments of 2. Mutation of the strategy involved flipping the bit value. Mutation of the tag involved replacing the real tag value with another uniformly randomly selected tag from the range [0..1]. To summarize, when an agent is selected for reproduction into the next generation, mutation is applied to the strategy bit (resulting in the bit being flipped with probability m) and to the tag (resulting in it being replaced with a new randomly selected tag with probability mf).

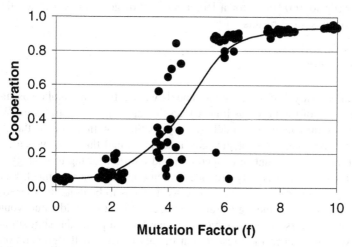

Fig. 1. Results from each simulation run plotting mutation factor (f) against cooperation

3.3.2 Results

The results are given in figure 1. Cooperation increases as the mutation factor is increased. For each value of the mutation factor (f) given on the x-axis are plotted 20 points from 20 individual runs (to 1000 generations). Cooperation given on the y-axis represents the proportion of all game interactions in a run that were mutually

---

[5] Using this method the probability that an agent will be reproduced into the next generation is probabilistically proportional to average payoff.

cooperative. Since we have 100 agents, with one game each per generation and 1000 generations per run, each point represents a proportion of mutual cooperation over 10^5 games. Each run had the same parameters but was initialized with different pseudo-random number seeds. The (smoothed) line joins the plotted average of the 20 points. The average is therefore over 2×10^6 individual games. To improve readability noise has been added to the x-coordinate of each point (+/-0.5). There are a number of interesting characteristics presented in figure 1. Firstly, we do indeed see an increase (on average) of cooperation when we increase the relative mutation rate of the tag with respect to the strategy. Given this we have a little more confidence that our hypothesis may be correct since it allowed us to predict this property.

The increase is non-linear, the average curve, appears, to approximate a sigmoid shape with three zones: A first zone with convergence to low value, a zone where it is unpredictable and a zone with convergence to high value. Where f < 4 we find convergence to low cooperation (no results above 0.2 cooperation[6]). For f > 6, cooperation converges to a high value (no results below 0.8 – note points that appear to violate this statement are a result of the added noise as mentioned above. In the "unpredictable area" $4 \leq f \leq 6$ we get high variance of results – indicating both high and low cooperation outcomes. Here, it would seem, results become unpredictable and chaotic (i.e. influenced by random variations due to the different pseudo-random number seed used in each run). When we ran the same experiments with larger agent populations (up to 1000) and for a larger number of generations (up to 10,000) we obtained broadly similar results.

4   Conclusions

From a detailed analysis of existing tag models we identified an implicit assumption – the mutation rate of the tags was higher than that applied to the strategies. We tested this hypothesis in a new tag model by varying the mutation rate of the tag while keeping the rate applied to strategies constant. We found that there was a non-linear relationship between amount of cooperation and the ratio of tag to strategy mutation rate. High cooperation was only produced when tag mutation was much higher than the strategy mutation rate. However, more work needs to be done in order to predict, for given scenarios, what the tag / strategy mutation ratio threshold value would be[7].

The results we present here are based on runs from a particular simulation model. However, we have (since the presentation of this paper and the preparation of this final publication draft) tested a number of models and found the general condition to hold [8, 9, 21, 22]. In addition, others have since confirmed our results [23].

The status of conclusions draw from empirical analysis of computer simulations is, as we have pointed out elsewhere [2], similar to those drawn from experimentation in the natural sciences. Without a deductive proof, results can always be challenged in the future by contradictory experiments or results from a sound deductive model. In

---

[6] Points that appear to violate this are a result of the added noise as mentioned previously.
[7] This will depend on a number of factors and a discussion is beyond the scope of, and space allowed for, this paper. See previous work [3, 4] for more on this.

this sense we tend to become more confident as more models reproduce the same or similar results but it should always be kept in mind that results based on simulation are not proofs.

We believe that the single-round PD potentially captures many kinds of engineering problem. *One area* we are currently exploring is a peer-to-peer engineering problem. If we can get nodes to cooperate in the PD then we believe we can use a similar technique to get them to share bandwidth and processing time, altruistically, in real systems. Recent work with network-like peer-to-peer simulation scenarios [8, 9] has shown that a high mutation rates on tags (or network links in these cases) are important in maintaining high cooperation and scalability. Our long-term aim is to produce a deployable service following a modular approach [12].

Acknowledgements

Thanks go to the anonymous reviewers who genuinely helped to make the paper a little more readable (all existing awkward phrasing and lack of clarity are of course the fault of the author who really should get his act together). Thanks go to Bruce Edmonds and Scott Moss from the CPM at Manchester Metropolitan University where the initial part of the work was supported. We are also grateful to Sendip Sen, Department of Mathematics & Computer Science, University of Tulsa, for sending a version of the paper presented at the AAAI Symposium [23].

References

1. Davis, L. (1991) Handbook of Genetic Algorithms. Van Nostrand Reinhold, New York.
2. Edmonds, B. and Hales, D. (2003) Replication, Replication and Replication - Some Hard Lessons from Model Alignment. Journal of Artificial Societies and Social Simulation 6(4).
3. Hales, D. (2000), Cooperation without Space or Memory: Tags, Groups and the Prisoner's Dilemma. In Moss, S., Davidsson, P. (Eds.) Multi-Agent-Based Simulation. Lecture Notes in Artificial Intelligence, 1979:157-166. Berlin: Springer-Verlag.
4. Hales, D. (2001) Tag Based Cooperation in Artificial Societies. PhD Thesis (Dept. Of Computer Science, University of Essex, U.K. 2001).
5. Hales, D. (2002) Evolving Specialisation, Altruism and Group-Level Optimisation Using Tags. In Sichman, J. S., Bousquet, F. Davidsson, P. (Eds.) Multi-Agent-Based Simulation II. Lecture Notes in Artificial Intelligence 2581:26-35 Berlin: Springer Verlag.
6. Hales, D. and Edmonds, B. (2003) Evolving Social Rationality for MAS using "Tags", In Rosenschein, J. S., et al. (eds.) Proceedings of the 2nd International Conference on Autonomous Agents and Multiagent Systems, Melbourne, July 2003 (AAMAS03), ACM Press, 497-503
7. Hales, D. and Edmonds, B. (2004) Can Tags Build Working Systems? - From MABS to ESOA. In Di Marzo Serugendo, G.; Karageorgos, A.; Rana, O.F.; Zambonelli (eds.) Engineering Self-Organising Systems - Nature-Inspired Approaches to Software Engineering. Lecture Notes in Artificial Intelligence 2977, Springer, Berlin.
8. Hales, D. (2004b) Self-Organizing, Open and Cooperative P2P Societies – From Tags to Networks. Presented at the 2nd Workshop on Engineering Self-Organizing Applications (ESOA) at AAMAS 2004, July 2004, New York.

9. Hales, D. (2004c) From Selfish Nodes to Cooperative Networks - Emergent Link-based Incentives in Peer-to-Peer Networks. Proceedings of the Fourth IEEE International Conference on Peer-to-Peer Computing (p2p2004), held 25-27 August 2004, Zurich, Switzerland. IEEE Press.
10. Hamilton, W. D. (1964) The genetical evolution of social behaviours, I and II. J.Theor.Biol.7,1-52.
11. Holland, J. (1993) The Effect of Lables (Tags) on Social Interactions. Santa Fe Institute Working Paper 93-10-064. Santa Fe, NM.
12. Jelasity, M., Montresor,A., and Babaoglu, O. (2004) A modular paradigm for building self-organizing peer-to-peer applications. Proceedings of the 1st International Workshop on Engineering Self-Organising Applications (ESOA 2003), Springer.
13. Kalenka, S., and Jennings, N.R. (1999) Socially Responsible Decision Making by Autonomous Agents. Cognition, Agency and Rationality (eds. Korta, K., Sosa, E., Arrazola, X.) Kluwer 135-149.
14. Nowak, M. & May, R. (1992) Evolutionary Games and Spatial Chaos. Nature, 359, 532-554.
15. Nowak, M. & Sigmund, K..(1998) Evolution of indirect reciprocity by image scoring. Nature, 393, 573-557.
16. Riolo, R. (1997) The Effects of Tag-Mediated Selection of Partners in Evolving Populations Playing the Iterated Prisoner's Dilemma. Santa Fe Institute Working Paper 97-02-016. Santa Fe, NM.
17. Riolo, R. L., Cohen, M. D. & Axelrod, R. (2001) Evolution of cooperation without reciprocity. Nature 414, 441-443
18. Roberts, G. & Sherratt, T. N. (2002) Nature 418, 449-500
19. Sigmund, K. and Nowak, A, M. (2001) Tides of Tolerance. Nature 414, 403-405.
20. Trivers, R. (1971) The evolution of reciprocal altruism. Q. Rev. Biol. 46, 35-57.
21. Hales, D. (in press) Understanding Tag Systems by Comparing Tag Models. Presented at the Second Model-to-Model Workshop (M2M2) co-located with the Second European Social Simulation Association Conference (ESSA'04) at Valladolid, Spain 16-19th of Sept 2004. Available at www.davidhales.com.
22. Edmonds, B & Hales, D. (in press) Computational Simulation as Theoretical Experiment. Journal of Mathematical Sociology. Available at www.davidhales.com.
23. McDonald, A and Sen, S. (in press) Analyzing the Effects of Tags on Promoting Cooperation in Prisoner's Dilemma. Presented at the AAAI 2004 Fall Symposium on Artificial Multi-agent Learning Symposium, October 21-24, 2004, Washington D.C.

Users Matter: A Multi-agent Systems Model of High Performance Computing Cluster Users

Michael J. North[1] and Cynthia S. Hood[2]

[1] Argonne National Laboratory, 9700 S. Cass Avenue, Argonne, IL
north@anl.gov
[2] Illinois Institute of Technology, Chicago, IL USA
hood@iit.edu

Abstract. High performance computing clusters have been a critical resource for computational science for over a decade and have more recently become integral to large-scale industrial analysis. Despite their well-specified components, the aggregate behavior of clusters is poorly understood. The difficulties arise from complicated interactions between cluster components during operation. These interactions have been studied by many researchers, some of whom have identified the need for holistic multi-scale modeling that simultaneously includes network level, operating system level, process level, and user level behaviors. Each of these levels presents its own modeling challenges, but the user level is the most complex due to the adaptability of human beings. In this vein, there are several major user modeling goals, namely descriptive modeling, predictive modeling and automated weakness discovery. This study shows how multi-agent techniques were used to simulate a large-scale computing cluster at each of these levels.

1 Introduction

High performance computing clusters (HPCC) have been a critical resource for computational science for over a decade and have more recently become integral to large-scale industrial analysis [1][2]. In addition, HPCC often provide the main processing capabilities for the emerging computational grids. In isolation, individual cluster components such as computing nodes, network cards and scheduling algorithms generally either perform as designed or have explainable failure modes. Despite both their long history and their well-specified components, the overall aggregate behavior of HPCC is poorly understood. The difficulties arise from complicated interactions between cluster components during actual operation.

The interactions within clusters have been studied by a variety of researchers using both real systems and models. These studies are reviewed in the following sections. While successful in many ways, the studies of real systems have been hindered by difficulties with properly controlling experiments and accurately observing the resulting metrics. The modeling research has also yielded substantial insights but has been limited to relatively simple cluster topologies or has focused on

P. Davidsson et al. (Eds.): MABS 2004, LNAI 3415, pp. 99–113, 2005.

major, but isolated, cluster subsystems such as an individual network layer or scheduler. Several researchers have identified the need for more comprehensive multi-scale modeling of high performance clusters. In particular, Downey and Feitelson have suggested the unrealized but "intriguing possibility" of models extensive enough to reach all the way to the user level and Dowdy et al. have called for a parsimonious and "holistic view" of HPCC [3][4]. This paper presents a multi-scale modeling framework that addresses these needs by supporting the simultaneous simulation of user process workloads, the Maui cluster scheduler and two network layers, namely Myricom Myrinet and Ethernet. Results from the simulation provide new evidence for substantial inefficiencies in the HPCC routing tables generated by the widely used Myrinet Mapper algorithm.

2 Related Work

A large number of studies of real HPCC systems have been completed in recent years. These studies complement the substantial body of existing work on related network and operating system issues. The direct study of HPPC systems has been successful in many ways. However, these studies have been hindered by difficulties with properly controlling experiments and difficulties accurately observing the resulting metrics. The examples reviewed below emphasize studies of the HPCC components that will be discussed later in this paper, namely Ethernet, Myrinet and backfilling schedulers.

Observational studies are analyses of existing data. These studies have covered topics ranging from transient job scheduling patterns to long-range correlations in user behavior. Representative examples include the observational studies completed by Feitelson and Nitzberg and Downey and Feitelson [5][3]. Observational studies of the type discussed above can provide definitive characterizations of selected HPCC performance issues. However, these studies are limited to available workloads and not all of the interesting variables can be observed.

Experimental studies are controlled comparisons of two or more distinct alternatives. Many experimental studies of real systems have been done covering topics ranging from transient job scheduling patterns to long-range correlations in user behavior. A representative example includes the experiments with backfilling performed by Feitelson [6]. As with observational studies, experimental studies are useful but difficulties in controlling cluster performance constrains the achievable experimental designs and not all of the interesting variables can be observed.

There exists a large body of work on the analytical modeling of HPCC and HPCC related issues. For example, Draper and Ghosh as well as others have created analytical models of wormhole routing. Kim and Lee created analytical models of Myrinet communication delays in a single Myrinet network switch [7][8]. These analytical studies have shed light on many critical features of cluster performance. However, these studies usually require simplifications such as limited topologies.

A large number of other simulations have been developed. For example, OPNET is a powerful commercial product that has many features for network and systems modeling. Chang reviews OPNET and compares it to a variety of other simulators [9]. Kang and

Cha built a simulation for evaluating the performance of Myrinet routers. Lawson and Smirni investigated multiple-queue backfilling scheduling using a model they developed. Many other simulators also exist [10][11]. Chang provides a survey [9].

3   The ClusterMod Framework

While successful for their purposes, the simulations discussed above do not include the full spectrum of HPCC interactions from the network protocol level all the way up to the cluster user level. Dowdy et al. have called for a parsimonious and "holistic view" while Downey and Feitelson have said the modeling all the way up to the user level is an "intriguing possibility" [4][3]. This is the focus of the ClusterMod framework.

3.1   Design Goals

All models are necessarily approximations. A given model can be used to answer a specific question if its approximations are consistent with the assumptions inherent in the question. A clear range of questions is therefore important for model design. The ClusterMod modeling framework has been developed to help address several general classes of questions. The motivation for these questions is detailed in the introduction and related work sections. The questions are as follows:

1. What factors influence high performance cluster process execution times? What effects do these factors have?
2. What performance results can be expected from a specific cluster structure executing chosen classes of workloads? What are the major causes and sources of variation?
3. How will alternative scheduling algorithms tend to behave given a specific cluster structure and chosen classes of work? Again, what are the major causes and sources of variation? Are there weaknesses in the scheduling algorithms that can be exploited by adaptive users?
4. How will alternative network routing protocols tend to behave given a specific cluster structure and chosen classes of work? As before, what are the major causes and sources of variation?
5. How will combinations of scheduling algorithms, network routing protocols, cluster structures, workloads and user behavior interact across the various implied scales of operation?

Naturally, these are broad questions. As was shown in the review of the related work, even beginning to answer these questions involves considerable amounts of work. The goal of the ClusterMod effort is not to immediately answer all of these questions. Neither is the goal to be a complete prepackaged solution for finding the answers. Rather, the goal is to provide an extensible framework that can be used by researchers investigating these questions. This paper presents the context of the framework, the framework itself and several example applications.

The ClusterMod modeling framework has been developed using a set of specific design goals that were derived from the questions detailed above. The goals help insure that models developed with ClusterMod can represent HPCC in sufficient detail to support meaningful research into the previously posed questions. The design goals are as follows:

1. The framework must support the modeling of network routing protocol behavior with enough detail to study communication link congestion and to investigate the effects of changes in protocol designs. This support must include the ability to model the simultaneous operation of several network routing protocols in a single cluster.

2. The framework must support the modeling of cluster structures with sufficient fidelity to investigate changes in performance that result from changing the availability or connectivity of network nodes and links.

3. The framework must support the modeling of user and system processes with enough detail to study the effects of different levels and kinds of workloads.

4. The framework must support the modeling of scheduling algorithms with sufficient detail to investigate the possible results from alternative scheduling approaches.

5. The framework must support the modeling of users with enough richness to study the potential consequences of cluster performance feedback and user adaptation.

6. The framework must weave together the modeling of network routing protocols, cluster structures, processes, cluster schedulers and users to allow the interactions between to be investigated.

These goals were used to develop and evaluate the model architecture. As such, ClusterMod provides researchers with the tools needed to implement models of both specific clusters and general cluster configurations. The use of ClusterMod to model Ethernet, Myrinet and the Maui Scheduler is discussed later in this paper. Furthermore, the use of these implementations models the Argonne Chiba City HPCC and to research the problem of process contention in HPCC is shown [12].

3.2 Software Architecture

ClusterMod models the components in an HPCC as a complex graph. The Cluster Node and Cluster Edge classes provide the foundation for ClusterMod HPCC graphs. Cluster Nodes maintain a list of incoming and a list of outgoing Cluster Edges as well as three-dimensional positioning information. Conversely, Cluster Edges maintain references to their terminal nodes. All individual edges in ClusterMod are unidirectional. Bidirectional links are modeled using two unidirectional edges, one in each direction. Edges that are naturally bidirectional, such as Ethernet links, take advantage of ClusterMod's automatic mechanisms for creating the two unidirectional edges needed when a single bidirectional edge is created.

Cluster Nodes are specialized into Units, Hosts, Management Units and Switches. Each of thee classes provide core methods and attributes that are used to model specific HPCC nodes. Unit is the parent class for Host and Switch. Host is the parent class for Management Unit.

Units naturally form the basis for the unit classes. They add NIC and ClusterMod Process management to Cluster Edges along with Maui Scheduler importing capabilities. The implementation of the Maui Scheduler in ClusterMod is discussed in a later section.

Cluster Processes provide the core components needed to model processes executing on Hosts. They are designed to model the central behaviors of host processes. Sets of Cluster Processes are grouped into Cluster Jobs that are scheduled by Schedulers.

Cluster Jobs are subclasses of Cluster Process that replace the normal Cluster Process compute cycle with process management behaviors. Jobs maintain a list of assigned units and processes. Jobs handle process initialization and monitor process termination once they are assigned resources by a Scheduler. Jobs are normally created by Cluster Users. Schedulers are specialized Jobs that manage other Jobs.

The central classes in the ClusterMod framework are shown using Unified Modeling Language (UML) notation in Figure 1 (left) along with an example ClusterMod Ethernet Network (right). The heart of ClusterMod is the Cluster Model. This component provides the central stage for the cluster modeling activities. There is one Cluster Model class instance per HPCC simulation. The Cluster Model maintains the discrete event time schedule, keeps a list of the active agents in the system, handles any required files and manages interactions with the model interface.

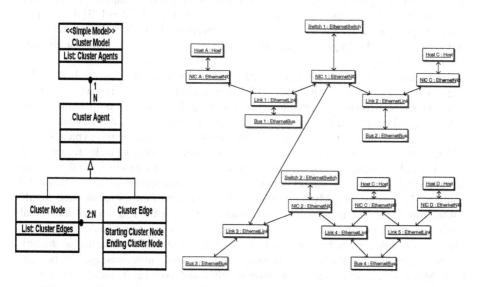

Fig. 1. The Core Classes (*left*) and an Example ClusterMod Ethernet Network (*right*)

The Cluster Model provides the central functions for all of the Cluster Agents used for simulation. The Cluster Agent class is also shown in Figure 1. Almost all of the customized classes used for modeling are derived from the Cluster Agent class.

Cluster Agent provides several key functions that are used throughout the ClusterMod framework. The Cluster Agent class provides an entrance to the time schedule, provides every component with a unique identifier, provides automatic storage capabilities for each component, and provides interactive user interface editors for each component. The time schedule functions are discussed in this section. The unique identifier (ID) is an integer that is used to track each component for a variety of different purposes including verification and validation.

Nearly all HPCC components are time-limited, asynchronous resources suggests that time tracking and nonbinding schedule access is an important abstraction that should be placed high in the ClusterMod component hierarchy. To factor this abstraction, the Cluster Agent class tracks component availability throughout each simulation run. This component availability is updated based on task times generated by a uniform random distribution that varies between a given minimum and maximum delay. These delay times are specified in seconds on a component-by-component basis, most commonly during component creation. However, these values can be changed at any time during a simulation to model component degradation, repair, or significant changes in state.

3.3   Implementation

Consistent with its design goals, ClusterMod was implemented using the Repast multi-agent simulation (MAS) toolkit. Repast is one of several advanced MAS toolkits that are currently available. For many other good examples, see both the survey by Serenko and Detlor and the survey by Gilbert and Bankes [13][14]. Repast is a free open source toolkit that was originally developed by Collier, Howe and North and others [15]. The Repast system, including the source code, is available directly from the web [16].

The Repast discrete event time schedule provides a convenient mechanism for tracking the global flow of time. This allows each of the ClusterMod HPCC components to schedule events for themselves and for other components in an organized and coordinated fashion. Nearly all of the simulation classes in ClusterMod are used to model HPCC components with time-limited availability. Typically, real HPCC components can only be used for one or a small number of tasks at any given time. Consider "phits," or physical units, on network links for example. In the case of an Ethernet link, only one packet can be meaningfully transmitted at a time without a collision. In the case of a Myrinet link, several flits can be transmitted at once. While considering these examples, it is important to observe that the activities of individual HPCC components are often asynchronous and vary in duration. Thus, an Ethernet link can transmit a message while network interface cards (NICs) attached to the link are both listening for the current message and preparing new messages to be sent later. Myrinet goes even further by having NICs and switches simultaneously send and receive messages passing through them from their links. On a large scale, this kind of implicit parallelism is one of the principle advantages of cluster systems. For example, the Myrinet link implementation uses the Cluster Agent's dynamic asynchronous scheduling capability to model the difference in time delays for the initial flit and the later flits in a message burst. A uniform distribution was chosen as the default since it is a simple random number source that can be used to approximate a reasonable range of behaviors, since it is fully specified by only two parameters and since it is naturally bounded from both above and below. Bounding from both above and below is particularly important since most of the processes to be modeled in HPCC take finite and strictly positive amounts of time. Other distributions can be substituted as needed. There are many choices including the common normal distribution, the triangular distribution and the general gamma distribution. See Law and Kelton for an extensive discussion of the alternatives [17].

Each time a simulated component needs something to occur, it calls one of the schedule methods to asynchronously queue the activity for future execution. There are three major types of schedule methods that scheduling events after waiting for a randomly distributed time period, schedule events when the agent is next available for work and schedule the event immediately. The types vary based on when the queued activity will occur. There are multiple versions of each of the types to support the queuing differing numbers of arguments to the called component. All of these methods extend the Repast schedule to add new functionality specifically for ClusterMod.

3.4 Verification

According to Law and Kelton, model verification is matching an implemented model to its design while model validation is matching an implemented model to the real world [17]. The ClusterMod main components were verified and validated using a combination of manual traces, extensive unit tests logged with Aspects and benchmark results [18]. An overview of Aspects can be found in Elrad, Filman and Bader [19].

The manual traces were used as part of the class verification and for interface verification. The traces involved the manual creation of small test cases that here subsequently simulated. The simulation cases involved one or two units and one link. The interface test cases involved from one to ten units and zero to twenty links.

The unit tests used Aspects to log calls to all of the core simulation functions [18][19]. The Aspects were used to insert logging code into all of the simulation methods [18][19]. This type of tracing covered all of the core simulation classes.

4 Myricom Myrinet, Ethernet and the Maui Scheduler

From Boden et al., Myrinet is a commercial network protocol created by Myricom, Inc. in the 1990's based on work done at the California Institute of Technology and the University of Southern California. Myrinet is a source-routed system that uses wormhole routing for flow control [20]. Myrinet was modeled in ClusterMod following the details given by Boden et al., Kim and Lee and other sources [20][8]. The ClusterMod Myrinet model was verified and validated using a combination of manual traces, extensive unit tests logged with Aspects and benchmark comparisons [18][19]. See the system verification and validation section for a background discussion.

An Ethernet model was also built using the ClusterMod framework. The results are similar to the Myrinet work described above. The ClusterMod Ethernet model was verified and validated using steps similar to that for the Myrinet model. The results of the ClusterMod Ethernet model were also compared to Francis, Frost and Soldan's measured Ethernet performance for multiple large file transfers as well as Smith and Kain's measurements of Ethernet performance under actual loads [21][22].

The Maui Scheduler is a free, open source cluster preemptive backfill job scheduler currently maintained as a community project [23]. The Maui Scheduler acts as an advanced HPCC resource manager. Maui was modeled in two different ways using the ClusterMod framework. The first approach modeled the full scheduler with the goal of supporting multi-scale investigations into Maui performance. The second approach modeled the behavior of known workloads with the goal of providing an experimental control for comparative studies. Both implementations followed the Maui Scheduler source code and other sources [23][24]. As with the previous models, the ClusterMod Maui Scheduler model was verified and validated using a combination of manual traces, extensive unit tests logged with Aspects and benchmark comparisons [18][19]. The system verification and validation section has a background discussion on this topic. The manual traces were used as part of the class verification. The unit tests used Aspects to log calls to all of the Maui Scheduler simulation functions [18][19]. These tests verified that the Full Maui Scheduler manages jobs as designed.

5   Users in ClusterMod

Many user studies have been completed as reviewed in the opening sections. However, despite this substantial body of work, much more remains to be done. In particular, Downey and Feitelson have emphasized the need for user modeling that goes well beyond the existing research [3]. Feitelson and Nitzberg and others have shown that the variable behavior of users has a large impact on cluster performance [5].

Dowdy et al. have noted that, not surprisingly, "users behave mischievously in order to beat the primitive scheduler, steal computational cycles and weasel in ahead of other users in the waiting queues" [4]. The Maui Scheduler Administrator's Guide states that the need for a specific scheduler feature "is necessitated by the fact that most sites have pretty smart users and pretty smart users like to work the system, whatever system it happens to be" [25]. The guide goes on to describe work done within the Maui Scheduler to control this kind of user behavior. Clearly this is not the last cluster or even scheduler feature that will be driven by user adaptation and learning. To address these issues, users need to be included in HPCC models.

There are several major user modeling goals, namely descriptive modeling, predictive modeling and automated weakness discovery. Current techniques for meeting these goals have met with significant success when applied other areas of research.

Descriptive user modeling summarizes or reproduces past user behaviors. It is the underlying data source for the other approaches as well as a productive line of research in its own right. The research examples have been cited earlier are all useful contributions, but again, according to Downey and Feitelson, more is needed [3]. In other areas, MAS techniques have been used successfully to describe competitive economic behavior similar to that of cluster users. North and Murakami et al. are two examples [26][27].

Predictive user modeling seeks to forecast future user behaviors. Explaining, predicting and budgeting runtime charges are immediate HPCC applications. There have been successes with other systems see Gozzi, Paolucci and Boccalatte and Veselka et al. for specific examples as well as Bonabeau's survey [28][29][[30].

Automated weakness discovery seeks to identify possible pathological behaviors. This line of research is differentiated from predictive studies because there is no attempt to determine whether users will actually take advantage of the discovered weaknesses. Rather, the goal is to determine if exploitable weaknesses exist. Related work in neighboring areas has met with significant success. See Ebben, de Boer and Pop Sitar and North, Macal and Campbell for examples of probing complex economic systems for weakness as well as platforms for such work [31][32]. The central focus for HPCC is the development and use of tools to probe cluster systems for potential weaknesses that might be exploited by users.

The descriptive user modeling done with the current ClusterMod system is not intended in any way to be a complete model of user behavior. It is only intended to indicate the potential of future research. The ClusterMod Scripted User was combined with the ClusterMod Ethernet, Myrinet and scripted Maui Scheduler classes to descriptively model user behavior recorded during May of 2001. Scripted Users run jobs based on predetermined plans. In this case, the plans were taken from reformatted logs for Argonne's Chiba City high performance computing cluster [12]. The correctness of the Scripted User behavior was confirmed using both manual and logged unit tests following the methods detailed in the system verification and validation section.

6 Software Development Methodology

ClusterMod was developed using object-oriented design with UML support as described in Booch [33]. Design patterns were used following Gamma et al. [34]. Aspects were used extensively for testing the framework. See Elrad, Filman, and Bader for an overview of Aspects [19]. ClusterMod was developed using the a variety of tools. The core agent –based modeling used Repast 2.1 [16]. The Javasoft Java 2 Development Kit, as described by Foxwell, was used for implementing ClusterMod [35]. Eclipse was used as the development environment. Freeman-Benson and Borning discuss the development of an urban simulation using Eclipse [36]. AspectJ was used for Aspect development following Walker, Baniassad, and Murphy [37]. The data captured with Aspects was logged with Log4j following Gülcü [38]. The use of Log4j, among other logging tools, in conjunction with AspectJ is discussed briefly by Cloyer et al. [18]. Unit testing was performed with JUnit as outlined in Beck and Gamma [39]. The WinCVS implementation of the Concurrent Versions System (CVS) was used for source code and data file version control following Fogel and Bar [40]. ClusterMod's three-dimensional graphical editor was custom built for the ClusterMod framework using Auburn University's Visualizing Graphs with Java (VGJ) library [41].

Summary software metrics for were calculated for both the framework itself and the Ethernet, Myrinet, and Maui Scheduler models. In general, ClusterMod scores well on the metrics. The outliers on the far right end of the distributions are the Cluster Model class and some of the import methods. The Cluster Model class is detailed in the section on core architecture. The import methods are described in the Ethernet, Myrinet and Maui sections. They are good candidates for future factoring.

Chiba City configuration data for selected hours from May 13 to 14, 2001 was obtained from the HPC Workload-Resource Trace Repository and from the Argonne Chiba City Project [32][12]. This data was used to simulate Chiba City's operation from Sunday May 13, 2001 at about 5:19 PM to Monday May 14, 2001 at about 4:30 A.M. Central Standard Time. The data described the hosts, the Myrinet network structure and the parameter settings for the Maui Scheduler as well as the requested workloads. The runs used the previously described full Maui Scheduler to reproduce the selected time. Three jobs were submitted during the simulation. These were the only jobs executed by the cluster during the simulation time window.

7   Initial ClusterMod Results

The ClusterMod Ethernet, Myrinet and Maui Scheduler classes described in previous sections were used to simulate Argonne's Chiba City high performance computing cluster [12]. Chiba City is a 256-host Linux cluster that is used for HPCC scalability research. The cluster is organized into "towns" of approximately 32 hosts each. Eight computing towns provide user services along with several supporting towns that provide management, storage and visualization facilities. Town activities are coordinated by the Maui Scheduler.

Chiba City configuration data for selected hours from May 13 to 14, 2001 was obtained from the HPC Workload-Resource Trace Repository and from the Argonne Chiba City Project [32][12]. This data was used to simulate Chiba City's operation from Sunday May 13, 2001 at about 5:19 PM to Monday May 14, 2001 at about 4:30 A.M. Central Standard Time. The data described the hosts, the Myrinet network structure and the parameter settings for the Maui Scheduler as well as the requested workloads. The runs used the previously described full Maui Scheduler to reproduce the selected time. Three jobs were submitted during the simulation. These were the only jobs executed by the cluster during the simulation time window.

The current model runtime for a group of large processes utilizing most Chiba City hosts is approximately ten to fifteen minutes for one hour of simulation time on an 2.08 GHz AMD Athlon XP 2800+ PC with 992 MB of RAM. Naturally, this run time is a function of the amount of activity in the cluster being simulated.

Figure 2 (left) shows the Myrinet structure of Chiba City on May 13 to 14, 2001 using the two-dimensional Fruchterman-Reingold (FR) graph layout algorithm [43]. The FR layout positions units based on their connectivity. It places heavily connected units such as switches near the center of diagrams and lightly connected units such as computational hosts towards the edges of diagrams. The Maui Scheduler is on the host in the lower left corner of the left diagram. Network traffic tends to move from far-flung hosts to the central switches and then back out again. The tree-like patterns created by this flow are visible in the diagram.

Figures 2 and 3 show normalized combined Myrinet usage for each of the three individual jobs that were submitted. Lines represent active Myrinet links. Note that all of the jobs are using a common set of links near the middle of the network. Also, notice that a few central switches are being heavily used by multiple jobs. These switches represent bottlenecks for Myrinet's wormhole routing protocol. Wormhole routing reservations keep nearly all of the downstream links utilized when key upstream switches become congested. This situation results in substantial system-wide congestion.

Figure 4 shows the logical process communications between the May 13 to 14, 2001 jobs. Each line represents one or more successful communications between processes on the connected hosts. The lines and nodes are shaded by user. The figure shows that the user processes were allocated disjoint sets of nodes with a significant amount of interleaving in Myrinet connectivity space. Comparing these results with those in the Figures 2 and 3 reveal that the Maui Scheduler's host allocation forced a few switches high in the network to be in the path between most links.

Why do the unit allocations appear to be fragmented in such a way as to increase network congestion when the Maui Scheduler is designed to allocate units in contiguous blocks based on host addresses? Results from the simulated Maui Scheduler show that the first job was allocated a single contiguous block of host addresses. However, as can be seen in Figures 2 and 3, these hosts were not always adjacent in the Myrinet network. The Maui Scheduler fits the second job into some of the spaces left over from the previous job's allocation. This, combined with the non-uniform Myrinet network structure, leads to the central switch congestion visible by comparing the conflicting switch usage in Figures 2 and 3. Finally, the situation worsened as the third job was interleaved into the spaces left behind by the other two, as can be seen by contrasting Figures 2 and 3. Key switches that are simultaneously demanded by two or more jobs are shown in the insets in the figures.

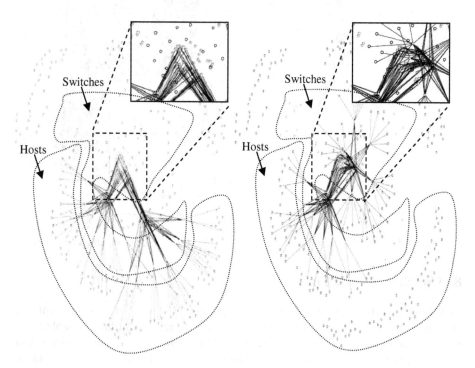

Fig. 2. User 1 Job 1 network usage (*left*) and User 1 Job 2 network usage (*left*)

The Chiba City Myrinet network is quite interleaved based on the switch host port connections tables from May 13 to 14, 2001. The simulation results suggest that allocating units contiguously by address under these conditions may result in heavily fragmented physical allocations and thus produce switch contention. Furthermore, the simulation results indicate that, at least for big jobs, the switch contention may occur in the critical upper-level switches. This can compound an already difficult situation.

As previously discussed, the Myrinet Mapper calculates routes within Myrinet networks. The Myrinet Mapper algorithm and the Maui Scheduler lead to the contention shown in Figures 2 and 3. This provides new evidence to support the concerns raised by Flich et al. and others that the widely used Myrinet Mapper algorithm can produce inefficient routes when it is employed in common cluster configurations [44]. Thus, the ClusterMod Chiba City simulation results help support calls for alternative Myrinet Mappers such as those proposed by Baik, Hood and Gropp and Flich et al. [45][44].

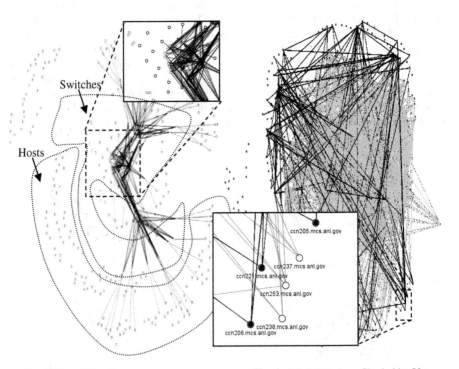

Fig. 3. User 2 Job 1 network usage   **Fig. 4.** Job Allocations Shaded by User

8   Conclusions and Future Work

Following Dowdy et al., a parsimonious and "holistic view" of high performance clusters is needed [4]. The extensive body of existing research, while valuable, does not adequately meet this need according to both Dowdy et al. and Downey and Feitelson [4][3]. Multi-scale high performance cluster simulation is needed to achieve the desired holistic view. The ClusterMod system provides a framework for multi-scale cluster simulation, as shown by its application to modeling Argonne's Chiba City high performance computing cluster. Results from this application provide new evidence for substantial inefficiencies in the HPCC routing tables produced by the widely used Myrinet Mapper algorithm.

There are several future directions for research with ClusterMod. First, ClusterMod will be used to investigate the factors that cause variations in cluster performance including network configurations, scheduler configurations, user requests and user process classes. Second, ClusterMod will be used to study the performance of alternate network protocols, such as those for message routing. Third, ClusterMod will be extended to study highly distributed heterogeneous grid computing clusters. Finally, ClusterMod will be extended to simulate adaptive users for prediction and automated weakness discovery. Each of these directions for future research is likely to improve our understanding of high performance computing clusters.

References

1. Dongarra, J., Meuer, H. Simon, H., and Strohmaier, E.: High Performance Computing Today, Proceedings of the 1st International Conference on Molecular Modeling and Simulation, American Institute of Chemical Engineers, New York, New York, USA (July 23-28, 2000) 1-7.

2. Sterling, T.: Launching Into the Future of Commodity Cluster Computing, Proceedings of the 2002 IEEE International Conference on Cluster Computing, IEEE, Piscataway, New Jersey, USA (Sept. 23-26, 2002) 345

3. Downey A., and Feitelson, D.: The Elusive Goal of Workload Characterization, ACM SIGMETRICS Performance Evaluation Review, Vol. 26 No.4, ACM, New York, New York, USA (Mar 1999) 14-29

4. Dowdy, L., Rosti, E., Serazzi, G., and Smirni, E.: Scheduling Issues in High-Performance Computing, ACM SIGMETRICS Performance Evaluation Review (Special Issue on Parallel Scheduling), ACM, New York, New York, USA (March 1999) 60-69

5. Feitelson, D. and Nitzberg, W.: Job Characteristics of A Production Parallel Scientific Workload on the NASA Ames iPSC/860, In , D. G. Feitelson and L. Rudolph (eds.): Job Scheduling Strategies for Parallel Processing, Lecture Notes in Computer Science, Vol. 949 Springer-Verlag, Heidelberg, Germany (1995) 337-360

6. Feitelson, D.: Experimental Analysis of the Root Causes of Performance Evaluation Results: A Backfilling Case Study, Technical Report 2002-4, School of Computer Science and Engineering, the Hebrew University of Jerusalem, Jerusalem, Israel (Mar. 2002)

7. Draper J., and Ghosh, J.: A Comprehensive Analytical Model for Wormhole Routing in Multicomputer Systems, Journal of Parallel and Distributed Computing, Vol. 23, No. 2, Elsevier, San Diego, California, USA (1994) 202-214

8. Kim, S., and Lee, S.: Measurement and Prediction of Communication Delays in Myrinet Networks, Journal of Parallel and Distributed Computing, Vol. 61, No. 2, Elsevier, San Diego, California, USA (2001) 1692-1704

9. Chang, X.: Network Simulations with OPNET, Winter Simulation Conference Proceedings, vol. 1, IEEE, Piscataway, New Jersey, USA (Dec. 5-8 1999) 307-314

10. Kang, S., and Cha, K.: Performance Evaluation and Simulation on Myrinet-Based Packet Router, Electrical Engineering 611 Class Project, School of Engineering, Cleveland State University, Cleveland, Ohio, USA (Fall 2000)

11. Lawson, B., and Smirni, E.: Multiple-Queue Backfilling Scheduling with Priorities and Reservations for Parallel Systems, ACM SIGMETRICS Performance Evaluation Review, Vol. 29, Issue 4, ACM, New York, New York, USA (March 2002) 40-47

12. Argonne National Laboratory, Chiba City Project: Available as http://www-unix.mcs.anl.gov/chiba/ (May 2004)

13. Serenko, A. and Detlor, B.: Agent Toolkits: A General Overview of the Market and An Assessment of Instructor Satisfaction With Utilizing Toolkits in the Classroom (Working Paper 455), McMaster University, Hamilton, Ontario, Canada (2002)

14. Gilbert, N., and Bankes, S.: Platforms and Methods for Agent-Based Modeling, Proceedings of the National Academy of Sciences of the USA, Vol. 99, Suppl. 3, National Academy of Sciences of the USA, Washington, DC, USA (May 14, 2002) 7197-7198

15. Collier, N., Howe, T., and North, M.: Onward and Upward: The Transition to Repast 2.0, Proceedings of the First Annual North American Association for Computational Social and Organizational Science Conference, Electronic Proceedings, Pittsburgh, PA USA (June 2003)

16. ROAD: Repast 2.0, Available as http://repast.sourceforge.net/ (May 2004)

17. Law, A., A. and Kelton, D.: Simulation Modeling and Analysis, McGraw-Hill, New York, New York, USA (1982)

18. Cloyer, A., Clement, A., Bodkin, R., and Hugunin, J.: Practitioners Report: Using AspectJ for Component Integration in Middleware, Companion of the 18th Annual ACM SIGPLAN Conference on Object-Oriented Programming, Systems, Languages, and Applications, ACM, New York, New York, USA (Oct. 2003)

19. Elrad, T., Filman, R., and Bader, A.: Aspect-Oriented Programming: Introduction, Communications of the ACM, Vol. 44, Issue 10, ACM, New York, New York, USA (October 2001) 29-32

20. Boden, N., Cohen, D., Felderman, Kulawik, R., Seitz, A., Seizovic, C., and Su, J.: Myrinet: Aa Gigabit-Per-Second Local Area Network, IEEE Micro, Vol. 15 , Issue 1, IEEE, Piscataway, New Jersey, USA (Feb. 1995) 29-36

21. Francis, S., Frost, V. and Soldan, D.: Measured Ethernet Performance for Multiple Large File Transfers, Proceedings of the 14th Conference on Local Computer Networks, IEEE, Piscataway, New Jersey, USA (Oct. 10-12, 1989) 323-327

22. Smith, W., and Kain, R.: Ethernet Performance Under Actual and Simulated Loads, Proceedings of the 16th Conference on Local Computer Networks, IEEE, Piscataway, New Jersey, USA (Oct. 14-17 1991) 569-581

23. Supercluster Research and Development Group: Maui Source Code, Available as http://www.supercluster.org/downloads/maui/ (Jan. 2004)

24. Supercluster Research and Development Group: Maui Scheduler Administrator's Guide v.3.2, Cluster Resources, Covered Bridge Canyon, Utah, USA (2002)

25. North, M.: Towards Strength and Stability: Agent-Based Modeling of Infrastructure Markets, Social Science Computer Review, Sage Publications, Thousand Oaks, California, USA (Fall 2001) 307-323

26. Murakami, Y., Minami, K., Kawasoe, T., and Ishida, T.: Multi-Agent Simulation for Crisis Management, Proceedings of the 2002 IEEE Workshop on Knowledge Media Networking, IEEE, Piscataway, New Jersey, USA (July 10-12, 2002) 135-139

27. Gozzi, A., Paolucci, M. and Boccalatte, A.: A Multi-Agent Approach To Support Dynamic Scheduling Decisions, Proceedings of the Seventh International Symposium on Computers and Communications, IEEE, Piscataway, New Jersey, USA (July 1-4, 2002) 983-988

28. Veselka, T., Boyd, G., Conzelmann, G., Koritarov, V., Macal, C., North, M., Schoepfle, B., and Thimmapuram, P.: Simulating the Behavior of Electricity Markets With an Agent-Based Methodology: the Electricity Market Complex Adaptive System (EMCAS) Model, Proceedings of the 22nd International Association for Energy Economics International Conference, Published on CD-ROM, Vancouver, British Columbia, Canada (October 2002)

29. Bonabeau, E.: Agent-Based Modeling: Methods and Techniques for Simulating Human Systems Proceedings of the National Academy of Sciences of the USA, Vol. 99, Suppl. 3, National Academy of Sciences of the USA, Washington, DC, USA (May 14, 2002) 7280-7287

30. Ebben, M., De Boer, L., and Pop Sitar, C.: Multi-Agent Simulation of Purchasing Activities in Organizations, Proceedings of the 2002 Winter Simulation Conference, Vol. 2, IEEE, Piscataway, New Jersey, USA (Dec. 8-11, 2002) 1337-1344

31. North, M., Macal, C., and Campbell, A.: Oh Behave! Problem Solving Environments for Agent Behavioral Simulation, International Journal of Future Generation Computer Systems, Elsevier, San Diego, California, USA (Accepted Jan. 2004)

32. Supercluster Research and Development Group: HPC Workload/Resource Trace Repository, Available as http://www.supercluster.org/research/traces/index.shtml (May 2004)

33. Booch, G.: Object-oriented Design with Applications 2nd ed., Addison-Wesley, Boston, Massachusetts, USA (1993)

34. Gamma, E., Helm, R., Johnson, R., and Vlissides, J.: Design Patterns: Elements of Reusable Object-Oriented Software, Addison-Wesley, Reading, Massachusetts, USA (1995)

35. Foxwell, H.: Java 2 Software Development Kit, Linux Journal, Specialized Systems Consultants, Seattle, Washington, USA (Oct. 1999)

36. Freeman-Benson, B., and Borning, A.: Practitioners Report: Experience in Developing the UrbanSim System: Tools and Processes, Companion of the 18th Annual ACM SIGPLAN Conference on Object-Oriented Programming, Systems, Languages, and Applications, ACM, New York, New York (Oct. 2003)

37. Walker, R., Baniassad, E., and G. Murphy: An Initial Assessment of Aspect-oriented Programming, Proceedings of the 1999 International Conference on Software Engineering, IEEE, Piscataway, New Jersey, USA (May 16-22, 1999) 120-130

38. Gülcü, C.: Log4j Delivers Control Over Logging, Java World, Online Magazine Available as http://www.javaworld.com/, IDG, San Francisco, California, USA (Nov. 2000)

39. Beck, K., and Gamma E.: Test Infected: Programmers Love Writing Tests, Java Report, vol. 3, issue 7, 101 Communications, Chatsworth, California, USA (1998) 37-50

40. Fogel K., and Bar, M.: Open Source Development with CVS, 2nd ed., Coriolis, Scottsdale, Arizona, USA (2000)

41. Barowski, L.: Visualizing Graphs with Java Library, Auburn University, Available as http://www.eng.auburn.edu/department/cse/research/graph_drawing/graph_drawing.html (Jan. 2004)

42. Fruchterman, T., and Reingold, E.: Graph Drawing by Force Directed Placement, Journal of Software: Practice and Experience, Vol. 21, No. 11, Wiley, New York, New York, USA (1991) 129-1164

43. Flich, J., Malumbres, M., Lopez, P., and Duato, J.: Improving Routing Performance in Myrinet Networks, Proceedings of the 14th International Parallel and Distributed Processing Symposium, IEEE, Piscataway, New Jersey, USA (May 1-5, 2000) 27-32

44. Baik, S., Hood, C., and Gropp, W.: Prototype of AM3: Active Mapper and Monitoring Module for the Myrinet Environment, Proceedings of the 27th Annual IEEE Conference on High Speed Local Networks, IEEE, Piscataway, New Jersey, USA (Nov. 6-8, 2002) 703-707

Formal Analysis of Meeting Protocols

Catholijn M. Jonker[a,1], Martijn Schut[a], Jan Treur[a,b], and Pınar Yolum[a,2]

[a] Vrije Universiteit Amsterdam, Department of Artificial Intelligence,
De Boelelaan 1081a, 1081 HV Amsterdam, The Netherlands
{jonker, schut, treur, pyolum}@few.vu.nl
[b] Universiteit Utrecht, Department of Philosophy,
Heidelberglaan 8, 3584 CS Utrecht, The Netherlands

Abstract. Organizations depend on regular meetings to carry out their everyday tasks. When carried out successfully, meetings offer a common medium for participants to exchange ideas and make decisions. However, many meetings suffer from unfocused discussions or irrelevant dialogues. Within Social Science sometimes general, informal meeting guidelines are formulated. To study meetings in detail, we first formalize general properties for meetings and a generic meeting protocol for the role interactions in meetings that is coherent with such guidelines. In the context of a case study, an example meeting is simulated based on this protocol. The properties are verified in this simulated trace. These properties are also validated by verifying them against a formalisation of empirical data of a real meeting in the same context. A comparison of the two traces reveals that a real meeting is more robust in the sense that exception violations of the protocol may occur, and these exceptions are handled effectively without damaging the success of the meeting. Given this observation, a more refined protocol is specified that includes exception-handling strategies. Based on this refined protocol a meeting is simulated that closely resembles the real meeting.

1 Introduction

Meetings are an integral part of every day life. Meetings are important tools in most organizations to structure decision processes and to disseminate information throughout the organization. Typically the members of a group come together on a regular basis to inform each other of new developments, to discuss problems, and propose solutions. While many organizations depend on face-to-face meetings, it is notoriously difficult to hold a focused and an effective meeting. There is an abundant literature on guidelines on how to carry a successful meeting [7, 1]. These guidelines are rather informal, which makes it hard to put into practice and hard to evaluate.

This paper formalizes a domain-independent meeting protocol that can be used in various meetings. The formalization captures many intuitive ideas that are also

---

[1] Currently at: Division of Cognitive Engineering, NICI, Nijmegen, The Netherlands, C.Jonker@nici.ru.nl.
[2] Currently at: Department of Computer Engineering, Bogazici University, Bebek, Istanbul, Turkey, pyolum@cmpe.boun.edu.tr

P. Davidsson et al. (Eds.): MABS 2004, LNAI 3415, pp. 114–129, 2005.

mentioned in meeting guidelines, hence is compatible with most meeting guidelines. The formalization captures actions that need to be carried out by participants as well as constraints that each participant has to satisfy. The main aim of this work is to understand how meeting protocols are carried out, by understanding the different flows that take place in meetings. To achieve this, we study the meeting protocol with an empirical trace as well as with a simulated trace and analyze various properties. The empirical trace is based on observations of a real meeting. The simulated trace is generated in a simulation environment where agents are assumed to follow the meeting protocol strictly. We compare the two traces in terms of desirable properties. Desirability is defined from the perspective of the attendees; it is the desire to let the meeting go as quickly as possible, to be fair to all attendees (chance to speak if they want to), necessary items are discussed, etcetera.

The rest of this paper is organized as follows. Section 2 develops the formal generic meeting protocol. Section 3 studies a generic meeting trace based on the formalized protocols. Section 4 introduces an empirical trace of a real meeting. Section 5 analyzes both traces formally in terms of desired properties. Section 6 provides a revised protocol and a simulation of the enhanced protocol. Section 7 discusses the relevant literature in comparison to this work.

2 Meetings Formalized

In this section a formalization of the organization of a meeting is presented: organizational structure (Section 2.1), dynamic properties for the overall process (Section 2.2.1), and a protocol for role interactions (Section 2.2.2).

2.1 Organizational Structure

Consider a typical meeting that contains a chairperson, a secretary, and a number of participants. A common form to structure meetings is the following. A Chairperson chairs every meeting. The Secretary takes minutes of the meeting. Taking minutes means writing down the arguments presented by the Participants of the meeting, as well as the decisions made. Chairing a meeting means opening and closing a meeting, making sure that people are talking one at a time, and that only the current issue is discussed. The decision process differs according to the customs and/or agreements in the group. Common decision procedures are decision by consensus, decision by majority, and decision by the Chairperson. A question to be addressed is how dynamic properties describing such a protocol can be identified.

2.2 Organizational Behavior

Dynamic properties characterizing an organizational behavior can be specified at different levels: at the level of the organization as a whole, at the level of interactions between roles (interaction protocol), and at the level of roles. We define an organization property as a characteristic that is exhibited by an organization as a whole. Such a property holds in an organization as a result of the individuals. The organization in this case is the meeting as a whole.

2.2.1 Organizational Behavior Properties

For the organisation of a meeting a number of organization properties can be identified. As an example the following property expresses that no two participants speak at the same time. In this and the following properties, communicates_from_to(p, q, x, y) denotes that p communicates to q the communicative act x with the content y. For this paper, we consider two types of communicative acts, mainly inform and declare. Only when the communicative act x is a "declare" act, then the receiver q is dropped meaning that the message is sent to everyone. For the sake of simplicity, we assume that messages always reach their destination.   For an explanation of the formal language TTL used, see [3,4].

Organizational Property 1 (OP1)

Informal

During the meeting only one Participant is speaking at a time.

Semiformal

At any point in time,

if   any participant is speaking,

then   all other participants are not speaking

Formal

$\forall t, p, p'$:PARTICIPANT, q, q' :ROLE, x, x', y, y'

$p \neq p'$ & state(γ, t, output(p)) |= communicates_from_to(p, q, x, y) $\Rightarrow$
state(γ, t, output(p')) |$\neq$ communicates_from_to(p', q', x', y')

To express the properties the following abstractions have been introduced for agenda item, current agenda item and addressed agenda item.

Abstraction: agenda item

Informal

An *agenda item* is an item that was declared to be an agenda item and not retracted since then

Semiformal

Item i is an agenda item if at some point in time it was declared to be so,

and since then it was not declared that it is no agenda item

Formal

agenda_item_at(γ, i, t) =

$\exists$ m:CHAIR, t' $\leq$ t

state(γ, t', output(m)) |= communicates_from_to(m, declare, agenda_item(i)) &
$\forall$t"   t' < t"< t state(γ, t", output(m)) |$\neq$ communicates_from_to(m, declare, not_agenda_item(i))

Abstraction: current agenda item

Informal

A *current agenda item* is one that was opened but not yet closed.

Semiformal

An agenda item is a current item if and only if

Some time ago the Chairperson declared that item to be the current item

And since then the Chairperson did not declare the item closed.

Formal

current_agenda_item_at(γ, i, t) =

∃m:CHAIR, t' ≤ t

state(γ, t', output(m)) |= communicates_from_to(m, declare, opened(i)) &

∀t" [t' < t"< t ⇒ state(γ, t", output(m)) |≠ communicates_from_to(m, declare, closed(i))]

Abstraction: addressed agenda item
Informal

An agenda item has been *addressed* if it was opened and closed during the meeting.

Semiformal

An agenda item has been addressed if and only if
for every time point that the chairperson has opened the item, at a later time point she declared the item closed

Formal

addressed_agenda_item_at(γ, i, t) =

∃ m:CHAIR, t1≤t state(γ, t1, output(m)) |= communicates_from_to(m, declare, opened(i)) &

∀t2≤t state(γ, t2, output(m)) |= communicates_from_to(m, declare, opened(i))
⇒ ∃t3 t2≤t3≤t &

state(γ, t3, output(m)) |= communicates_from_to(m, declare, closed(i))

OP2
Informal

During the meeting only agenda items are addressed.

Semiformal

At any point in time t,
if   the item i is opened
then   i is an agenda item

Formal

∀t, i, p, q, x, y ∀m:CHAIR

state(γ, t, output(m)) |= communicates_from_to(m, declare, opened(i))

⇒ agenda_item_at(γ, i, t)

OP3
Informal

Every Participant who indicates that he has something to say on the current agenda item will have the opportunity to speak.

Semiformal

At any point in time t,
if   a participant communicates that he has something to say about the current agenda item i
then   before the item was closed a later time point exists such that at t' the participant communicates something in the context of i

Formal

∀t, I, p:PARTICIPANT, q:ROLE current_agenda_item_at(γ, i, t)
&

state(γ, t, output(p)) |= communicates_from_to(p, q, inform, has_input_for(p, i))
⇒

∃t' ≥ t, x
state(γ, t', output(p)) |= communicates_from_to(p, q, inform,x)
& is_in_context_of(x, i)

The notion of being in context of is assumed to be a given notion.

OP4
Informal
Eventually the meeting is closed.
Semiformal
At some point in time the chairperson declares the meeting closed
Formal
∀m:CHAIR ∃t state(γ, t, output(m)) |= communicates_from_to(m, declare, meeting_closed)

OP5
Informal
If the meeting is closed, all agenda items have been addressed.
Semiformal
At any point in time,
if   the meeting is declared closed,
then for any item i that was on the agenda there are earlier time points at which item i was declared opened and closed
Formal
∀t, i, m:CHAIR
    state(γ, t, output(m)) |= communicates_from_to(m, declare,
    meeting_closed) & agenda_item_at(γ, i, t) ⇒
addressed_agenda_item_at(γ, i, t)

OP6
Informal
No two items are current at the same time.
Semiformal
At any point in time t,
if   item i is current at t,
and   item i' is current at t,
then   i = i'
Formal
∀t, i, i'
current_agenda_item_at(γ, i, t) & current_agenda_item_at(γ, i', t) ⇒ i = i'

OP7
Informal
If a participant is speaking, then she is speaking on the current item.
Semiformal
At any point in time t,
if   at t the item i is current agenda item
 and   at t any participant is communicating X,
then   X fits in item i

Formal

∀t, i, p, q :ROLE, x, y
current_agenda_item_at(γ, i, t) & state(γ, t, output(p)) |= communicates_
from_to(p, q, x, y)
⇒ is_in_context(y, i)

OP8

Informal

The meeting starts and ends in time.

Semiformal

The meeting starts at the planned starting time
and ends before the planned end time

Formal

∀m:CHAIR ∀t1 [planned_starting_time(t) ⇒
state(γ,   t1,   output(m)) |=   communicates_from_to(m,   declare,
meeting_opened))] &
∀t2, t3 state(γ, t2, output(m)) |=   communicates_from_to(m, declare,
planned_end_time(t3)) ⇒
∃t4 ≤ t3   state(γ, t4, output(m)) |= communicates_from_to(m, declare,
meeting_closed))

OP9

Informal

Every communication in the meeting is received by everyone

Semiformal

At any point in time,
if   a participant communicates something to another one,
then   this communication will be received by everyone

Formal

∀t, p, q , q':ROLE, x, y
state(γ, t, output(p)) |= communicates_from_to(p, q, x, y) ⇒
∃t'≥t   state(γ, t, input(q')) |= communicates_from_to(p, q', x, y)

OP10

Informal

The secretary will make minutes of the meeting

Semiformal

if   an agenda item is closed,
then   notes for the minutes on this item have been made by the Secretary

Formal

∀t, i ∀m:CHAIR
state(γ, t, output(m)) |= communicates_from_to(m, declare, closed(i)) ⇒
state(γ, t, EW) |= notes_present_for_by(i, Secretary)

OP11

Informal

The internal state property of the chairperson indicating that i is being discussed
holds precisely then when i is a current agenda item

Semiformal

if   an agenda item is closed,

then   notes for the minutes on this item have been made by the Secretary

Formal

∀t, i ∀m:CHAIR

current_agenda_item_at(γ, i, t) ⇔

state(γ, t, internal(m)) |= being_discussed(i)

2.2.2 Role Interaction Properties: The Generic Meeting Protocol

A number of role interaction properties have been specified to define a generic interaction protocol for a meeting. Here role interaction properties can also be seen as constraints on agents' interactions.

RI1   If the Chairperson generates a question (which implies a permission to speak) to a Participant, then a little time later the Participant generates an answer.

Formal

∀m:CHAIR, p:PARTICIPANT ∀t

[state(γ, t, output(m)) |= communicates_from_to(m, p, request, q)) &

not ∃ x state(γ, t', output(p)) |= communicates_from_to(p, m, inform, x)]

∃t' > t   state(γ, t', output(p)) |= communicates_from_to(p, m, inform, answer_on(a, q)))

RI2   If a Participant requests to add an item to the agenda,
               then the Chairperson communicates this to all Participants.

Formal

∀m:CHAIR, p:PARTICIPANT ∀t

state(γ, t, output(p)) |= communicates_from_to(p, m, request, agenda_item(i)))

⇒

∃t' > t   state(γ, t', output(m)) |= communicates_from_to(m, declare, agenda_item(i)))

Notice that it is not difficult to express in these properties within how many seconds a reaction should be given. For simplicity this has been left out.

RI3   If the Chairperson generates a permission to speak for a Participant,
               then that Participant will begin speaking on the current agenda item.

Formal

∀m:CHAIR, p:PARTICIPANT ∀t

state(γ, t, output(m)) |= communicates_from_to(m, p, permit, speak)) &

current_agenda_item(i)

⇒ ∃t' > t, y

state(γ, t', output(p)) |= communicates_from_to(p, m, inform, y) &

is_in_context_of(y, i)

RI4   If the Chairperson revokes the permission to speak from a Participant while that Participant is still speaking, then that Participant will stop speaking immediately.

Formal

∀t, i ∀m:CHAIR, p:PARTICIPANT

state(γ, t, output(m)) |= communicates_from_to(m, p, revoke, i)

& state(γ, t, output(p)) |= communicates_from_to(p, x, y, i)

⇒ ∀x', y', z' state(γ, t+1, output(p)) |≠ communicates_from_to(p, x', y', z')

RI5   If all Participants who at an earlier point in time have indicated that they have information or a question regarding the current item, have put forward their information,

then the Chairperson asks each Participant in turn whether he has further information on the current item.

Formal

∀t, i ∀m:CHAIR

[∀p:PARTICIPANT [∃t''≤t state(γ, t'', output(p)) |= communicates_from_to(p, m, inform, has_input_for(i)) ⇒

∃t'''≤t, y state(γ, t''', output(p)) |= communicates_from_to(p, m, inform, y) & is_in_context _of(y, i)]]

⇒   ∃ t'≥t   state(γ, t, output(m)) |=   communicates_from_to(m, request, further_info_on(i))

RI6   If the Chairperson has declared an agenda item closed, and not all items have been treated, then the Chairperson will announce one of the remaining items as the current item.

Formal

∀t, i ∀m:CHAIR

state(γ, t, output(m)) |= communicates_from_to(m, declare, meeting_closed)

& ∃i agenda_item_at(γ, i, t) & not addressed_agenda_item_at(γ, i, t)

⇒ ∃i, t'≥t agenda_item_at(γ, i, t) & not addressed_agenda_item_at(γ, i, t) & state(γ, t', output(m)) |= communicates_from_to(m, declare, opened(i))

RI7   If the Chairperson has declared the meeting opened,

then the Chairperson will announce the proposed end time.

Formal

∀t, i ∀m:CHAIR

state(γ, t, output(m)) |= communicates_from_to(m, declare, meeting_opened)

⇒ ∃t'≥t, t''   state(γ, t', output(m)) |=   communicates_from_to(m, declare, planned_end_time(t''))

RI8   If the Chairperson has proposed an end time,

then the Chairperson will announce the agenda items.

∀t, t'', i ∀m:CHAIR

state(γ, t, output(m)) |= communicates_from_to(m, declare, planned_ end_time(t'')) &

agenda_item_at(γ, i, t)

$\Rightarrow$ $\exists t' \geq t$ state(γ, t, output(m)) |= communicates_from_to(m, declare, agenda_item(i))

RI9   If the Chairperson has announced all agenda items,
    then the Chairperson will ask if any Participant has another agenda item.

$\forall t, t'', i\ \forall m$:CHAIR
$\forall i$ [agenda_item_at(γ, i, t) $\Rightarrow$
$\exists t' <$ t   state(γ, t', output(m)) |= communicates_from_to(m, declare, agenda_item(i))]
$\Rightarrow$ $\exists t'' \geq t$ state(γ, t'', output(m)) |= communicates_from_to(m, request, other_items)) &

RI10   If the Chairperson has declared the last agenda item closed and 10sec has
    passed,
    then the Chairperson will close the meeting.

$\forall t, t'', i\ \forall m$:CHAIR, $\forall i$ [agenda_item_at(γ, i, t) $\Rightarrow$
$\exists t' \leq t$ state(γ, t', output(m)) |= communicates_from_to(m, declare, closed(i))]
&
$\exists i$ [agenda_item_at(γ, i, t) &
state(γ, t, output(m)) |= communicates_from_to(m, declare, closed(i))] &
$\Rightarrow$ $\exists t'' \geq t$ t'' < t + 10
state(γ, t'', output(m)) |=   communicates_from_to(m, declare,
    meeting_closed))

RI11   If all Participants have answered that they have no further information on the
    current item,
    then the Chairperson provides a summary and declares the item closed.

Formal
$\forall t, i\ \forall m$:CHAIR
[$\forall p$:PARTICIPANT state(γ, t, output(p)) |= communicates_from_to(p, m, inform, no_further_info_on(i))]
$\Rightarrow$ $\exists\ t' \geq t$ state(γ, t, output(m)) |= communicates_from_to(m, declare, summary(i)) &
state(γ, t, output(m)) |= communicates_from_to(m, declare, closed(i))

3 Simulating a Meeting Based on the Generic Meeting Protocol

The simulations of interest are generated using a logic-based simulation environment. Using this environment, executable temporal rules are specified so that the simulation environment can generate a trace, for more details see [3]. These executable temporal rules are executed based on the current status of the world, without regard to the past. A generated trace describes which state properties related to the protocol hold at each time point. The generated traces can then be analyzed with an automated logic-based checker. This checker takes as input a property of interest about the trace and logically validates the property by the trace. If the property holds in the trace, the checker outputs 'success' otherwise it outputs 'fail'.

We consider a simulation of an example meeting on the topic of study groups. These simulations consist of one chairperson (referred to as chair) and three participants (referred to as p1, p2, and p3). The agenda items are about particular study groups, hence named as group_1, group_2, and so on. For each of the agenda items one of the participants is the contact person, who is asked to speak if the agenda item is opened.

Simulation as discussed here is based on the formal specification of the generic meeting protocol, which was developed based on the meeting guidelines discussed above. The simulation follows the protocol but here we give a brief overview of the trace. The simulation starts by the chairperson declaring the desired end time (proposed_end_time) for the meeting. Next, the chairperson announces the agenda items one by one (agenda_item). Next, the chairperson asks for further additions to the agenda. Participant p1 suggests a new item (schedule), which is also added to the agenda. Once the agenda is finalized, the chair opens the first item (group_1) for discussions. The chairperson requests information from the participant who is likely to have input on the current agenda item. After this participant is done speaking, the chairperson asks the other participants to see if they have further information for the topic (last_comments). Since no participant has further input on the agenda item (group_1), the chairperson closes the agenda item and opens the second item. This procedure repeats itself until the agenda item is group_4. On this agenda item, when the chairperson asks for other comments from the participants, participant p3 provides additional comments. Later the meeting is continued as before. After the last agenda item is discussed, the chairperson declares the meeting closed. A complete trace can be found in [3].

From a broad overview, the simulation described above has some differences from our observations of real meetings. For this reason, we observed a real meeting and obtained data on how it was carried out. These data were analyzed in some depth.

4 An Empirical Trace of a Real Meeting

An important part of the work presented here is based on empirical data. This data was obtained through carefully observing a meeting in the Artificial Intelligence Department of the Vrije Universiteit Amsterdam. Similar to the observation techniques explained elsewhere [6], the observer sat apart from the meeting participants and the chair. Two of the participants and the chair knew why the observant was present, while a third participant did not.

The observer wrote down the conversations of the meeting in an informal language. Later these informal texts were formalized to analyze and reason about the meeting. Table 1 gives brief snapshots from this. For a complete formalized trace see [3]. The left column in the table provides the informal text and the right column gives the formalized states.

We briefly explain the differences from the simulated meeting trace in Section 5. The trace again starts with the chairperson announcing a desired end time for the meeting (proposed_end_time). The chairperson announces the agenda items but does not explicitly ask for additions to the agenda. After the chair opens an agenda item and receives input on the item, she closes the item when she sees fit. Compared to the generic meeting protocol described in Section 2.2.2, the difference here is that the chair does not explicitly ask for further input from the participants. Complementing this is a change in the role behavior of participants. Whereas in the meeting simulated according to the generic protocol (Section 3), a participant speaks only when permission is given, in the real meeting participants take the initiative to speak up without being asked. The interesting question then is how these different behaviors affect the outcome of the meetings? Do the desired properties of interest hold for both cases? Does one trace have advantages over the other one? We discuss these questions next.

Table 1. The transition from informal statements to formal states

	Informal Description	Formal State
...		
2	C: We will talk about the regular agenda	communicates_from_to(chair, declare, agenda_item(group_1)) communicates_from_to(chair, declare, agenda_item(group_2)) communicates_from_to(chair, declare, agenda_item(group_3)) communicates_from_to(chair, declare, agenda_item(group_4)) communicates_from_to(chair, declare, agenda_item(group_5))
...		
8	C: Mike any inputs for group_2	communicates_from_to(chair, p1, request, group_2)
9	Mike gives an explanation on group_2	communicates_from_to(p1, chair, inform, group_2)
10	Mike complains about lecture notes	communicates_from_to(p1, chair, inform, notes)
11	C: This is not the right time for that.	communicates_from_to(chair, p1, revoke, notes)
12	C: Let's move on	communicates_from_to(chair, declare, close(group_2))
...		
20	C: Let's move on	communicates_from_to(chair, declare, close(group_4))
21	C: Group_5	communicates_from_to(chair, declare, open(group_5))
22	C: Tim, any inputs for group_5	communicates_from_to(chair, p2, request, group_5)
23	Tim speaks more on group_4	communicates_from_to(p3, chair, inform, group_4) communicates_from_to(chair, declare, open(group_4))
24	C: We talked enough on group_4	communicates_from_to(chair, p3, revoke, group_4) communicates_from_to(chair, declare, close(group_4))
25	C: Group_5	communicates_from_to(chair, declare, open(group_5))
...		
32	C: OK, we are done now.	communicates_from_to(chair, declare, close(schedule))
33	C: Same time, next week	communicates_from_to(chair, declare, meeting_closed)

5   Formal Analysis of Simulated Trace and Empirical Trace

We analyzed the traces generated by these simulations in terms of the organization properties defined above, for more properties see [3]. To do so, the organization properties of Section 2 (and more) have been entered into the checker and automatically checked against each trace.

5.1 Analysis of the Simulated Meeting

The meeting simulated according to the generic protocol (Sections 2 and 3) satisfies the first organization property (OP1) which states that no two participants speak at the same time. This is intuitive since participants speak only when given permission. In this simulation, the chair ensures that only one participant has the permission to speak. Hence, the property holds. The second property (OP2) is on the agenda items that were talked. The role interaction RI6 specifies that once an agenda item is closed, then the chair chooses a new item from the agenda. Hence, it is always the case that the chairperson will open an existing agenda item. This explains why OP2 holds for this trace as well.

OP3 is satisfied for this trace because before closing each topic the chairperson asks for further comments from the participants. Hence, anyone who declares an intention to speak will get a change to speak. Organization property OP4 states that the meeting is eventually closed. This will always hold for a meeting based on the generic meeting protocol as long as the number of items on the agenda as well as the duration of comments on the items is finite. OP5 ensures that no meeting ends prematurely; that is if the meeting ends, then all agenda items have been discussed. In the specification of the meeting, the only way to close a meeting is when the meeting items have been discussed. OP6 states that no two items are open at the same time. This holds for this trace again due to role interaction RI6. A chairperson will open a new agenda item only if the previous item is closed. Organization property OP7 states that if a participant is speaking then she is speaking on the current item. This follows from the fact that the chairperson will only allow a participant to speak on the current item (RI3). Organization property (OP8) states that meeting start and end on time. This property holds for this trace since the first thing in the traces there is a declaration of intended start and end times of the meeting and that the meeting takes place between these time points. However, in general this property may have conflicts with OP3.

5.2 Analysis of the Empirical Data of the Real Meeting

While the generic meeting protocol obediently obeys the organization properties, the real meeting trace violates some of them. To avoid repetition, only the properties that are violated are discussed here.

The first interesting situation happens during the discussion of item group_3. The chairperson requests information from p2 on the item. The participant p2 speaks with short breaks (stammer), which influences one of the other participants (p3) to help p2 with his speech (complete). Notice that this is not part of the generic protocol and in general no participant has to help other participants. To be able to generate this behavior, we added an extra role interaction property to the simulation so that participant p3 would help p2. Participant p3's helping p2 is constructive in that it allows p2 to formulate his thoughts. Ironically, this situation disobeys one of the desired organization properties of meetings; namely OP1 which states that no two participants at a meeting should speak at the same time.

After a chair person requests information from a participant, the participant provides the required information. In some cases, it could also be the case that the participant provides information that is not relevant to the request of the chairperson. One such example happens during the discussion of item group_2 (see lines 7-12). After giving feedback on group_2, participant p1 starts speaking on a topic (notes) that is out of the scope of group_2. This is an example of impromptu interruption from participants that sometimes happen. This behavior of p1 causes the violation of the organization property OP7, which says that participant speak on current agenda items only. While this behavior of the participant is not part of the generic interaction protocol, a method for recovering from such a situation is followed in the meeting. Hence, the chair person can first revoke the permission from participant p1 and then continue with the protocol.

Contrary to the generic protocol, in this simulation the chairperson does not request further input from other participants before closing an agenda item. One interesting consequence is that after the discussion of item group_4, the chairperson closes the agenda item (line 20). However, there is still a participant who is willing to speak more on the item. Hence, this participant (participant p3) continues speaking about group_4, even though the item has been closed and a new item has been open (line 23). This point in time is interesting because in reality both agenda items are current. Item group_5 is current because it has been declared as open and not closed by the chairperson. While group_4 is also current, since one participant is talking about this item. Hence, another organization property, property OP6 is violated since there are two current items at the same time. However, this failing of this property does not halt the system. The meeting handles this exception in the sense that the chair person in this case lets the participant finish and then re-closes the item group_4 and reopens the item group_5 (in lines 24 and 25).

6 Refined Protocol and Simulation

As shown in the analysis in Section 5, a real meeting (such as the one described in Section 3) may deviate from a meeting correctly following the protocol (such as the simulated meeting in Section 4) in the following ways:

- sometimes, by exception, protocol properties are violated by one of the members
- strategies are employed to handle these exceptions and get the meeting on the right track again

One of the reasons that these exceptions occur is the fact that human agents are not ideal and may forget things. In practice members are able to accept these shortcomings and to recover from them. To this end a number of exception handling strategies are used. This can be considered a more sophisticated way of working than just by following the protocol. An interesting question is whether the generic meeting protocol can be refined by including such exception handling strategies to provide a more robust protocol. This question is discussed in the current section.

To experiment with a refined protocol, using the formal states given for the empirical trace, a second simulation was developed, where a number of the properties for the simulation (as used in Section 3) were adapted to reconstruct the empirical trace as precisely as possible. The generated trace indeed closely resembles our observations of the real meeting described in Section 4. For example, the exception of the participant speaking on notes while the current agenda item is group_2, is now handled realistically in the simulation: the chairperson first revokes the permission from participant p1 and then continues with the protocol. Moreover, now also the simulated meeting can handle the exception that during an item i1 a participant wants to add to an already closed agenda item i2. The strategy was added that for such an exception the chairperson returns to the earlier agenda item i2, lets the participant finish and then re-closes the item i2 and reopens the item i1. The following properties, that can be considered part of such a refined protocol, were used to obtain this:

RI1 If after a new agenda item was opened and not yet closed, a Participant speaks on an earlier addressed agenda item,
  then the Chairperson closes the current agenda item and reopens the earlier item.

Formal

$\forall$t, i1,i2 $\forall$m:CHAIR, p:PARTICIPANT
[current_agenda_item_at((γ, i2, t) &
addressed_agenda_item_at(γ, i1, t) &
state(γ, t, output(p)) |= communicates_from_to(p, m, inform, y)   &
in_context_of(i1)]
$\Rightarrow$ $\exists$ t''≥t state(γ, t'', output(m)) |= communicates_from_to(m, declare, closed(i2)) &
state(γ, t'', output(m)) |= communicates_from_to(m, declare, opened(i1))

RI2 If a Participant speaks on an item other than the current agenda item or any earlier addressed agenda item,
  then the Chairperson revokes the Participant and asks for additional comments on the current agenda item from the other participants.

Formal

$\forall$t, i2 $\forall$m:CHAIR, p:PARTICIPANT, y
[state(γ, t, output(p)) |= communicates_from_to(p, m, inform, y) &
not $\exists$i1 addressed_agenda_item_at(γ, i1, t) & in_context_of(y, i1)]
$\Rightarrow$ $\exists$ t''≥t state(γ, t'', output(m)) |= communicates_from_to(m, p, permission, revoke) &
$\forall$q state(γ, t'', output(m)) |= communicates_from_to(m, q, request, info_on(i1))

Using these properties, a new trace was generated that shows how participants can accommodate these exceptions. More information on this trace can be found in [3].

7 Discussion

In this paper a generic role interaction protocol for meetings that adhere to several guidelines on holding meetings was formalized, using the logical language TTL; cf. [3]. Moreover, desirable overall properties for a meeting were formally specified. In a case study in terms of the desirable overall properties of a meeting, an empirical trace was compared with a simulated trace generated from the given meeting protocol. Based on deviations revealed in this comparison, a more human-like refined protocol was specified and used as a basis for another simulation, closely resembling the empirical data.

Croston and Goulding present one of the earlier empirical works on meeting effectiveness [2]. Croston and Goulding develop a meeting analysis kit that is used in different departments of a company by the participants of the meeting. The kit enables the participants to re-evaluate a past meeting by analyzing the discussed topics, the time spent on each topic, and so on. Based on the analysis from different meetings, Croston and Goulding observe that the starting a meeting with a formal agenda and better chairing of the meetings increase the effectiveness of meetings. The meeting protocol that we propose respects both of these observations. Further, we explicitly formalize the notion of better chairing a meeting.

Table 2. Simulation trace based on the refined protocol

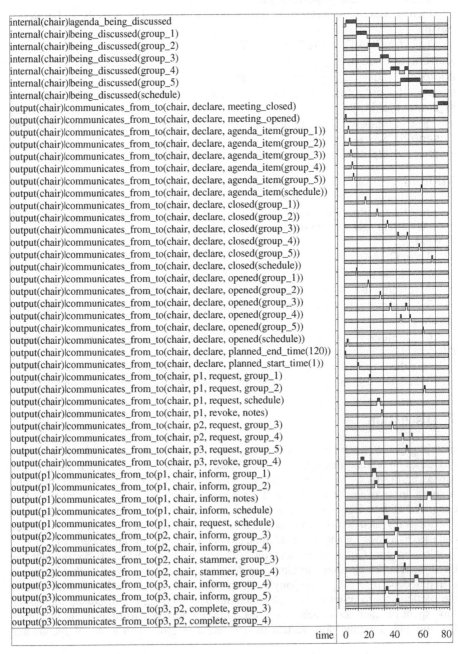

```
internal(chair)|agenda_being_discussed
internal(chair)|being_discussed(group_1)
internal(chair)|being_discussed(group_2)
internal(chair)|being_discussed(group_3)
internal(chair)|being_discussed(group_4)
internal(chair)|being_discussed(group_5)
internal(chair)|being_discussed(schedule)
output(chair)|communicates_from_to(chair, declare, meeting_closed)
output(chair)|communicates_from_to(chair, declare, meeting_opened)
output(chair)|communicates_from_to(chair, declare, agenda_item(group_1))
output(chair)|communicates_from_to(chair, declare, agenda_item(group_2))
output(chair)|communicates_from_to(chair, declare, agenda_item(group_3))
output(chair)|communicates_from_to(chair, declare, agenda_item(group_4))
output(chair)|communicates_from_to(chair, declare, agenda_item(group_5))
output(chair)|communicates_from_to(chair, declare, agenda_item(schedule))
output(chair)|communicates_from_to(chair, declare, closed(group_1))
output(chair)|communicates_from_to(chair, declare, closed(group_2))
output(chair)|communicates_from_to(chair, declare, closed(group_3))
output(chair)|communicates_from_to(chair, declare, closed(group_4))
output(chair)|communicates_from_to(chair, declare, closed(group_5))
output(chair)|communicates_from_to(chair, declare, closed(schedule))
output(chair)|communicates_from_to(chair, declare, opened(group_1))
output(chair)|communicates_from_to(chair, declare, opened(group_2))
output(chair)|communicates_from_to(chair, declare, opened(group_3))
output(chair)|communicates_from_to(chair, declare, opened(group_4))
output(chair)|communicates_from_to(chair, declare, opened(group_5))
output(chair)|communicates_from_to(chair, declare, opened(schedule))
output(chair)|communicates_from_to(chair, declare, planned_end_time(120))
output(chair)|communicates_from_to(chair, declare, planned_start_time(1))
output(chair)|communicates_from_to(chair, p1, request, group_1)
output(chair)|communicates_from_to(chair, p1, request, group_2)
output(chair)|communicates_from_to(chair, p1, request, schedule)
output(chair)|communicates_from_to(chair, p1, revoke, notes)
output(chair)|communicates_from_to(chair, p2, request, group_3)
output(chair)|communicates_from_to(chair, p2, request, group_4)
output(chair)|communicates_from_to(chair, p3, request, group_5)
output(chair)|communicates_from_to(chair, p3, revoke, group_4)
output(p1)|communicates_from_to(p1, chair, inform, group_1)
output(p1)|communicates_from_to(p1, chair, inform, group_2)
output(p1)|communicates_from_to(p1, chair, inform, notes)
output(p1)|communicates_from_to(p1, chair, inform, schedule)
output(p1)|communicates_from_to(p1, chair, request, schedule)
output(p2)|communicates_from_to(p2, chair, inform, group_3)
output(p2)|communicates_from_to(p2, chair, inform, group_4)
output(p2)|communicates_from_to(p2, chair, stammer, group_3)
output(p2)|communicates_from_to(p2, chair, stammer, group_4)
output(p3)|communicates_from_to(p3, chair, inform, group_4)
output(p3)|communicates_from_to(p3, chair, inform, group_5)
output(p3)|communicates_from_to(p3, p2, complete, group_3)
output(p3)|communicates_from_to(p3, p2, complete, group_4)
```

time   0   20   40   60   80

Serman and Basili study various properties of software inspection meetings in a software development project [6]. Although these types of meetings are different from the ones presented here, their and our procedures for meeting analysis have similarities. Similar to the generation of the empirical trace here, Serman and Basili collect data by attending inspection meetings as an observant. They later analyze their data statistically to uncover causal relations between various properties of the meeting, such as effectiveness, efficiency, or meeting length. While Serman and Basili discover interesting relations, they do not provide a formal protocol of how the meetings should be carried out as we have done here. Since our study uses simulations, we can easily adjust different behaviors of participants to see the effect of (local) properties of participants of a meeting on the (global) properties of the meeting as a whole.

Generally, the group-support systems help participants share data, improve communication, and reach decisions. Hence, group-support systems can help increase the efficiency of meetings. Niederman *et al.* study the meetings in organizations with group-support systems [5]. Their primary focus is to show how the use of group-support systems by facilitators affects meeting performances. Through interviews with facilitators, Niederman et al. observe that different facilitators have different ideas on measuring performance. However, no formal properties for identifying or bringing out successful meetings are identified.

Given the informal literature as discussed, the work reported in the current paper contributes some first steps in formal analysis of meetings. It is shown how meeting simulations following widely accepted guidelines in a rigid manner, do not resemble human meetings, which exploit more sophisticated strategies. It is pointed out how this discrepancy can be overcome by allowing by exception violations of the protocol, and by including exception handling mechanisms and strategies within the protocol. Future research will address this theme further.

References

1. James L. Creighton. Using Group Process Techniques to Improve Meeting Effectiveness. URL: http://www.effectivemeetings.com/teams/teamwork/creighton.asp
2. J. D. Croston and H. B. Goulding. The Effectiveness of Communication at Meetings: A Case Study. Operational Research Quarterly, Vol.17. No.1, pp. 47-57, March 1966.
3. Catholijn M. Jonker, Martijn Schut, Jan Treur, and Pinar Yolum. Formal Analysis of Meeting Protocols. Vrije Universiteit Amsterdam, Department of Artificial Intelligence. Technical Report, 2004. Available at: http://www.few.vu.nl/~wai/Papers/TR2004-meeting.pdf
4. Catholijn M. Jonker and Jan Treur. Compositional Verification of Multi-Agent Systems: a Formal Analysis of Pro-activeness and Reactiveness. International Journal of Cooperative Information Systems, vol. 11, 2002, pp. 51-92.
5. Fred Niederman, Catherine M. Beise, and Peggy M. Beranek. Issues and Concerns about Computer-Supported Meetings: The Facilitator's Perspective. MIS Quarterly, Vol.20, No. 1, pp. 1-22, March 1996.
6. Carolyn B. Serman and Victor R. Basili. Communication and Organization: An Empirical Study of Discussion in Inspection Meetings. In IEEE Transactions on Software Engineering, Vol.24, No. 6, pp. 559-572, July 1998.
7. Kevin Wolf. The Makings of a Good Meeting. October, 2002. Available at: http://members.dcn.org/kjwolf

From KISS to KIDS
– An 'Anti-simplistic' Modelling Approach

Bruce Edmonds and Scott Moss

Centre for Policy Modelling,
Manchester Metropolitan University
http://cfpm.org

Abstract. A new approach is suggested under the slogan "Keep it Descriptive Stupid" (KIDS) that encapsulates a trend in increasingly descriptive agent-based social simulation. The KIDS approach entails one starts with the simulation model that relates to the target phenomena in the most straight-forward way possible, taking into account the widest possible range of evidence, including anecdotal accounts and expert opinion. Simplification is only applied if and when the model and evidence justify this. This contrasts sharply with the KISS approach where one starts with the simplest possible model and only moves to a more complex one if forced to. An example multi-agent simulation of domestic water demand and social influence is described.

1 Introduction

The popular admonition to "Keep It Simple Stupid" or *KISS*, makes good sense if you are designing or constructing something – that is, if one has a particular specification, function or purpose in mind and one is trying to construct something appropriate. In this context the advice makes sense in two ways: *firstly*, that the more complex something is the less easy it is to control (and hence to make it do what one wants); and, *secondly*, in circumstances where a particular design does not work that one should resist elaborating it in an attempt to make it work but rather one should engage in a more fundamental re-evaluation (since such elaboration seldom really works) [7].

However the *KISS* sentiment is often also applied to the business of modelling some phenomena, e.g. with a simulation or a set of equations. So that even when faced with obviously complex phenomena modellers strive to keep their models simple, to an extent that is beyond any evident justification. The reasons for this are varied, if often left implicit. There are obvious practical reasons for keeping a model simple – this makes it easier to: implement; manipulate; analyse; check; communicate, etc.. However, it is sometimes claimed (and often implied) that the advantages of simplicity go beyond these practical values – that a simpler model is more likely to be true; or gets closer to the essence of the matter. In other words, that a simpler model is, in general, fundamentally better – it is this we are arguing against.

We suggest a new slogan: "Keep It Descriptive Stupid" or *KIDS* – which is supposed to suggest the approach to modelling where one starts with a descriptive model (which may be quite complex) and then only simplifies it where this turns out

P. Davidsson et al. (Eds.): MABS 2004, LNAI 3415, pp. 130–144, 2005.

to be justified. This is in contrast to the *KISS* paradigm where one *only* tries a more complex model if simpler ones turn out to be inadequate. Multi-agent based simulation (MABS) not only facilitates the *KIDS* approach but epitomises it. Thus the importance of the move towards MABS can be seen as part of a broader movement away from unjustified abstraction in modelling – abstraction that is, since the advent of accessible computational power, frequently unnecessary.

This can be seen as the encapsulation of a trend in some agent-based social simulation work where relatively rich models are being developed, often in close collaboration with relevant stakeholders. We do not have space here or the time for a survey of such models, but point to some recent work, including many in: the journal JASSS (http://jasss.soc.surrey.ac.uk); the recent ESSA (http://essa.eu.org) conference; and a forthcoming special issue of *Simulation* on applications of agent-base social simulation. Examples of work progressing in this direction include [2, 3, 15, 22, 23].

We necessarily can not deal with all the issues this paper raises – this would take a book rather than a workshop paper. Thus this paper has to be more in the way of a summary which passes over many of the arguments. We are forced to refer to our previous work for many of these issues – these will provide a more adequate entrance to the relevant literatures. We apologise for this. Thus for more on the meaning and definition of complexity see [6]; for the issue of how and why one might judge one model to be better that another see [9, 11]; for more on the relation of the scientific process to simplicity see [7]; for an extended paper on why simplicity is not truth-indicative see [10]; for more on the relation of formal systems to the scientific process see [8]; for more on the intended theory/implemented simulation distinction and its ramification see [9, 14]; and for more on constructive ways forward see [11].

2 Why *KIDS* Rather than *KISS*?

We are limited beings, which could explain why simplicity has such advantages and attractions for us. It also seems we have a tendency to project our own characteristics upon the natural world, thus we would like to *think* that everything has an "inner" simplicity even if they *appear* to be very complicated. Sometimes this is expressed as an assumption that simplicity is a (fallible) guide to truth, sometimes by conflating simplicity with generality (hiding the assumption that simpler models will apply to a greater variety of real cases, an assumption that is often unjustified).

When a modelling decision is justified using a phrase like "for the sake of simplicity" it is implicit that this is (in some sense) a good justification. People do not tend to justify their modelling decisions in papers using phrases such as: "it would have taken too long"; "we could not think how to do this"; or "it would make it very hard to check", despite the fact that, in our view, *these are perfectly acceptable* (indeed inevitable) reasons for beings with limited resources (like us). Rather, often it seems that people invoke simplicity because of its positive connotations, even if it was the more practical reasons that motivated them (of course, they also do so because that is what they were taught, but presumably the teachers had some reason for supposing this etc.). Thus the philosophical tradition that somehow simplicity is truth indicative gives credence to modelling decisions that are not otherwise explicitly justified. Elsewhere I (BE) argue that simplicity is *not* truth-indicative [10].

Rather we suggest that one would expect that, for many purposes and for many phenomena, that the models will need to be complex if they are to be adequate for the purposes for which the model is built. Given that much that we study is complex, it would be very surprising if it always turned out to be the case that the models could be simple. Surely the burden of proof is on those who insist that it is not sensible to try and match the complexity of the model with the complexity of the phenomena being modelled. The facts that complex outcomes *can* emerge from apparently simple systems (as in mathematics or ALife) does not mean that the complex phenomena we now observe is reducible to simple models. Even if (a huge presumption) a certain phenomena was generated using simple mechanisms, this does not mean the results we observe now are simple or could be usefully represented with a simple model. For more on the difference between kinds of complexity see [6].

To make this clear consider an analogy with a 19th Century naturalist making sketches of animals. The naturalist is doing this in order to help correctly identify and classify new species. It would have been laughable to suggest that such sketches should be limited to line cartoons "for the sake of simplicity", "so that they have the greatest possible generality", because the different species would not have been identified yet. It may have been that small details would have turned out to be important later for distinguishing closely-related species. Only when the naturalist had examined (and sketched) enough individuals would it have become clear which details were relevant and which not. The details that turned out to be essential may have allowed the drawings to be significantly simplified *a posterior*, but in many cases only marginally simplified (as with moths). The point is that it is simply not appropriate to make simplifications *before* one knows what is relevant and what is not. Simply hoping that one may have chanced upon an appropriate simplification does not make that simplification justified.

This is even more evident when one is considering the domain of interacting systems of flexible and autonomous actors or agents. The relevant behaviour of many such systems will not be simple, in particular they will not be reducible to aggregate models (for example statistical models) without significantly diverging from their target systems (that is, the agents are 'embedded' in their society in the sense of [17]). This includes multi-agent systems that are designed for a specific purpose [12]. Adopting a multi-agent model represents a move towards descriptive accuracy since: actors or agents in the target domain are represented by agents in the model and communications by messages between the agents. That is MABS allows and facilitates a more direct correspondence between what is observed and what is modelled. One benefit of this move to descriptive MABS is that a whole swath of evidence becomes available for validating our models – straight-forward descriptive evidence gained from observation of the domain can be applied to the model by virtue of the more straight-forward correspondence. What is new is that this evidence may be anecdotal or "common-sense". Previously such evidence may have been rejected on the ground that it is not "scientific" or "rigorous", but this was because it was not formalisable in terms of the current modelling technology (analytic mathematics) and hence had no deducible outcomes that could be checked. Now such qualitative information *can* be formally modelled in simulations where the deduction of outcomes is performed computationally rather than analytically. Further, in a descriptive MABS, this is relatively easy and natural and combines naturally with participative approaches to model construction and validation, as in [2].

Given that such evidence can be made rigorous and thus brought within the domain of the scientific, it should not be ignored. Thus if you have a target domain in which there is such evidence (such as "it seems that the actors learn to avoid areas where they were mugged") then models should only ignore such evidence if either: (1) there is good reason to think this is irrelevant or (2) there is evidence that this is wrong (e.g. over-simplified to the extent that it is significantly erroneous). In particular if one has access to a direct or expert "common-sense" account of a particular social or other agent-based system, then one needs to justify a model that ignores this solely on *a priori* grounds. In other words, constructing a model that is simpler *only* "for the sake of simplicity" *may* be a case of wilfully ignoring evidence.

Thus a move towards descriptive models allows for *more* of the available evidence to be applied. Of course, all available evidence should be used, including evidence that is traditionally used: time series data, point measurements, statistics etc. and evidence resulting from new methods of data collection (as suggested in [4]). This can be used as a sort of check of the anecdotal evidence, because one can check that a model constructed on the basis of anecdotal evidence is consistent with them (see [20] or references above for many others who suggest this). As with all data, one takes into account its reliability. Anecdotal evidence *can* be unreliable – however this does not mean we should not use it, merely that we need to cross-check it (as with any other useful but fallible evidence), but this is exactly what MABS facilitates.

This is an uncomfortable lesson for us: that the common-sense description may be a far better starting place than an artificially simple construction based on the guesses of academics. However, if one is modelling social or multi-agent systems, where the phenomena is undoubtedly complex (making *a priori* guessing difficult) it is not sensible to ignore any evidence that may help us. The more complex the phenomena the more evidence as to its workings are needed.

One common response to the arguments above is to claim that one is doing a sort of applicable mathematics rather than science. That is, one is not claiming that a particular model (or simulation) represents *in any sense* any observed phenomena but that one is merely establishing the model's properties, so that in the future someone may be able to successfully apply it to solving real problems. I call this the "formalist stance" – it is often adopted in fields which are currently lacking significant empirical or practical success (e.g. AI or Economics). The argument goes thus: much mathematics that was driven by goals other than representing reality (but sometimes including simplicity) turned out to be very useful later on, might not the same thing happen with developing abstract simulations. The answer is that this is possible, but then the simulation (or model) should be judged by the same sort of criteria as is used to judge new mathematics, namely: precision, soundness, importance and generality. The assumptions/mechanisms upon which the results depend should have been made completely explicit and precise. The model should be very thoroughly tested to almost eliminate any possibility of error in the results (this will almost certainly involve a wide search of the parameter space and independent replication in the case of complex simulations). The results should be sufficiently important in that they effectively present (to us humans) new information about the system that is not available from a casual inspection of the simulation set-up and inputs. Finally, it should be established that the results are applicable to a wide range of kinds of system. All too often simulations that are claim protection under the formalist stance do not justify themselves against these criteria but rather via the credibility of their results in terms of observed phenomena. Thus they fail to satisfy any set of relevant academic criteria. These issues are discussed further in [8].

To summarise, the *KISS* approach says that one needs a good reason not to use the simplest model, and then allows for progressively more complex models if simpler ones turn out to be inadequate. In contrast the *KIDS* approach starts with a model that is as straight-forwardly descriptive as evidence and resources allow (even if this means that one starts with relatively complex model) and then allows for progressive development later (including simplification and abstraction) as evidence and understanding of the model support this (it may turn out that some features that turn out to be important later have been left out, leading to an even more complex model). This is shown in Fig. 1.

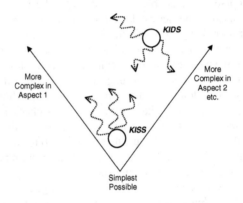

Fig. 1. An Illustration of *KISS* and *KIDS*

3 Distinguishing *Intended Theories* and *Implemented Models*

In this discussion it is important to distinguish between the intended theory and the implemented simulation (or model) that is meant to realise that theory in computational (or analytic) form. There are many ways in which this distinction is significant. In general, there will be many ways of constructing a simulation to embody any particular theory, because only certain aspects of the simulation are considered significant in terms of the theory, the rest being considered as the mechanisms to produce these. There will be various artefacts introduced as a side-effect of a particular implementation, so that the implemented model differs from the intended theory. This might be in ways that are considered unimportant, for example in minuscule deviations due to the pseudo-random number generator, or it may turn out to be significant (when a seemingly-innocent implementation detail significantly changes the results, e.g. as documents in [21]). All too often the implemented simulation and the intended theory are conflated or the intended theory is not described explicitly. This distinction is discussed further in [13].

In this paper we are arguing that the intended theory should (as a starting point) be as descriptive as possible and that a (single) simulation should be as direct a representation of this theory as is feasible. Of course, if there are two completely adequate ways of implementing the same theory as simulations then (for practical reasons) it is sensible to choose the simpler one. However feasibility considerations concerning the implemented simulation should not lead one to simplify the intended theory – this would be a case of "the tail wagging the dog". Rather if it is infeasible (for whatever reason) to implement a simulation of the intended theory, then this should be explicitly acknowledged.

When it is infeasible to implement a simulation that is adequate (for a specific purpose) to the intended theory (a common case in social simulation) one has a number of ways in which one can proceed. Often the wisest course is to abandon formal (including computational) modelling as (at least currently) beyond our capabilities. This may be followed by a decision to change the modelling goal to something less ambitious – for example to restrict the theory to cover a small aspect or special case. This might result in a new intended theory that is feasible to implement. The temptation here is to not admit one has taken this step – to pretend (to oneself) that the resulting feasible simulation is somehow still a model of the original intended theory.

4 Building Upwards from Descriptively Adequate Models

In this section we briefly outline a suggestion as to a constructive way forward in the face of extremely complex phenomena (such as with social systems and 'non-toy' multi-agent systems). More about this can be found in [11]. This suggestion is encapsulated in the diagram in Fig. 2.

Before we even get to descriptive simulations one has a series of "Data Models" [23] – that is, descriptions or data obtained by either measurement or elicitation. These provide the foundations upon which descriptive simulations will be built.

When one has a number of descriptive simulations concerning a related class of phenomena (satisfying similar purposes), one may be in a position to see what is relevant to a more abstract simulation, by examining the processes that result in these. It may well be that the abstract simulation only applies to the behaviour of descriptive simulations under certain conditions or (to use terminology from physics) within certain *phases*.

It may then be possible to formulate analytic models about the behaviour of aspects of the abstract simulation. For example, it may be observed that a certain process dominates the results in certain conditions and this can be approximated by some analytic or statistical models. The huge advantage of approaching abstract and analytic models in this way is that it provides a traceable chain of reference to observations of the phenomena. Thus if we ask why a certain term has a particular exponent in an analytic model, we can point to the behaviour in the abstract simulation this approximates. This can guide us when adapting or improving the analytics, especially when we are fault finding. Similarly, the mechanisms in the abstract model can be traced to the descriptive models it represents etc.

If one is in the fortunate position where a number of these analytic models or abstract simulations concur then this might be a sought-after general theory covering a range of phenomena. Such a theory can then be tested against the class of phenomena it concerns.

Later on, once the general theory has been thoroughly validated in multiple ways, it may be possible to simplify and systematise it. This process has occurred frequently in Physics. Whereas only a few people could understand Newtonian mechanics as it was first presented, nowadays it has been systematised into the accounts found in standard school text books. Such systemisation and simplification greatly facilities its applications to problems and new phenomena.

It is highly unlikely that 'short-cuts' to the simple or general theories will be discovered without a substantial amount of lower level data-collection and modelling has occurred first. This seems to be because the human mind requires conceptual frameworks to work within, which then trap us into formulating models similar to those that have gone before. It often seems to require a substantial (if sometimes indirect) 'jolt' from the phenomena itself to guide us towards really useful theory.

To summarise this, complex phenomena will not only require more complex simulations, but also the development and maintenance of complex *clusters* of models, as suggested in [16]. There are other ways in which such clusters may be created, such a using a number of models in parallel to cover different aspects of the same intended theory, this is discussed to a limited extent in [11].

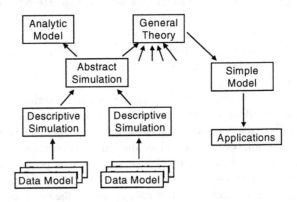

Fig. 2. An illustration of the bottom-up way of model development

5   Exploring Variations of a Model of Domestic Water Demand

The purpose of this section is to illustrate the *KIDS* approach. Thus we have here started with a model that encapsulates the aspects of the target domain for which we have (formal or informal) information. We then explore the behaviour of the model when different aspects stay the same – that is to see if one can find any aspects which do *not* have a relevant and significant impact on the chosen outcomes. This would then suggest a hypothesis that a simpler model is possible, which excludes this aspect. If all aspects seem to be essential to maintaining the properties of the outcomes that are deemed to be representative of what is reported, then this suggests that a simpler model would not be descriptively adequate. If this is the case it is difficult to see that such a model would have been reached by starting from a simpler model and elaborating it. Of course what we *have* is a complex model whose behaviour is not fully understood – it acts as an intermediary between observation and theory building.

The model in this example is a descriptive social simulation. It seeks to see how the patterns of domestic water demand in localities may be explained by mutual influence. To be exact, it models how a set of stakeholders *perceived* that households might interact because it was developed as the result of feedback from a panel of representatives from UK water companies and other domain experts (i.e. we would explain the model and current results and they would comment upon it). This model aimed to: capture their qualitative informal suggestions (e.g. demand rebounds fairly fast after a drought); be consistent with known data about households (e.g. ownership/frequency/use data of appliances); and have aggregate demand patterns similar to those observed (e.g. with clustered volatility)[1]. In other words, where we evidence, even if it was of an anecdotal nature, we used it in the design of the model. Only where we did not have any evidence did we turn to theory, and then theory developed as the result of at least some observation. For example, in this model although we had some evidence that households did influence their neighbours as to water usage the exact nature of the imitation behaviour was not known, which is why we turned to sociology to fill the gap. It is this insistence that the all the *available* evidence be reflected in the model design and behaviour, that makes it "descriptive".

The model was developed as part of the FIRMA[2] and CC:DEW[3] projects. FIRMA was an EU 5FP project to investigate the use of agent-based simulation and Integrated Assessment to fresh-water management issues in 5 river basins in Europe. CC:DEW was a project commissioned by the Environment Agency in the UK to assess the impact of any climate change on the demand for water. For a more detailed description of the model see [5]. The initial model was written by Scott Moss and then developed by Olivier Bathelemy and Bruce Edmonds. Tom Downing gave substantial advice and guidance.

The core of this model is a set of agents, each representing a household, which are situated on a grid. Each of these households is allocated a set of water-using devices in a similar distribution to those in the mid-Thames region of the UK. At the beginning of each month each household sets the frequency the appliance is used (and in some cases the volume at each use, depending on the appliance). Households are influenced as to their usage of these appliances by several sources: their neighbours and particularly the neighbour most similar to themselves (for publicly observable appliances); the policy agent; what they themselves did in the past; and occasionally the new kinds appliances that are available (in this case power showers, or water-saving washing machines). The individual household's demands are summed to give the aggregate demand. Each month the ground water saturation is calculated based on weather data (which is past data or past simulated data), if this is less than a critical amount for more than a month, this triggers the policy agent to suggest a lower usage of water. If a period of drought continues it progressively suggests using less and less water. The households are biased to attend to the influence of neighbours or the policy

---

[1] The validation of this model was quite complex and multi-faceted. For more details on this see [20].

[2] http://firma.cfpm.org

[3] http://www.sei.se/oxford/ccdew

agent to different extents – the proportion of these biases are set by the simulator. This structure is illustrated in Fig. 3.

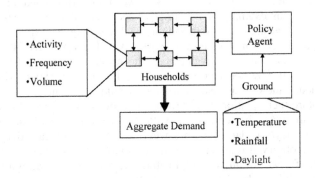

Fig. 3. The structure of the water demand model

The neighbours in this model are those either those in the shape of a cross or a square neighbourhood (Fig. 4). The extent of this neighbourhood is parameterised by the area (in the cases in Fig. 4 this is 8).

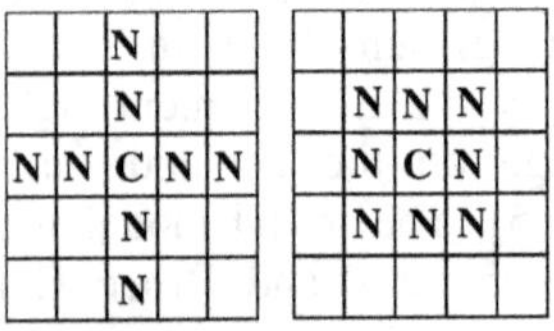

	N						
	N				N	N	N
N	N	C	N	N	N	C	N
	N				N	N	N
	N						

Fig. 4. The neighbourhood pattern for the households, left cross, right square (each area 8), C marks the focus location, and its neighbourhood is marked with Ns.

Every neighbour has a unique most neighbour who is most influential to it, the topology of this social network consists of a few pairs of mutually most influential neighbours and a tree of influence spreading out from these. The extent of the influence that is transmitted over any particular path of this network will depend upon the extent each node in the path is biased towards being influenced by neighbours.

Households are also (to a lesser extent) influenced by all its neighbours in its neighbourhood. The edges of this may or may not be wrapped around into a torus. The focus model used an unwrapped cross-shaped neighbourhood so the households at the edges and corners have fewer neighbours that those in the middle. The reason for this that the resulting patterns seem to us a reasonable mix of locality and complexity (data on the actual patterns was not available).

In each run the households are distributed and initialised randomly, whilst the overall distribution of the ownership and usages of appliances by the households and the biases of the households is approximately the same. In the runs described herein the same weather data is used, so the timing of droughts (and hence advice from the policy agent) and of new innovations are the same in each run.

The graphs below (such as Fig. 5 immediately below) show the aggregate demand resulting from many runs of the same set-up (rescaled so that 1973=100 for ease of comparison). Each line shows a different run from using that set-up, so you can see the variations in aggregate behaviour possible from the same model. Significant events include the droughts of 1976 and 1990, which often show up in a (temporarily) reduced water demand, due to agents taking the advice from the policy agent to use less water. Power showers become available in early 1988 and water-saving washing machines in late 1992 which can cause a sudden increase or decrease respectively.

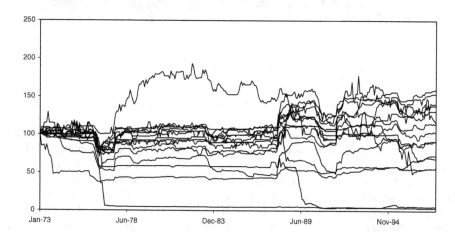

Fig. 5. Relative aggregate demand levels (cross-shaped unwrapped neighbourhoods of size 24)

Fig. 5 shows the model set-up chosen as the starting point for the model variations. It shows 15 different runs of that model. The droughts of 1976 and 1990 are clear as dips in demand in most (but not all) of the runs and the introduction of power showers in 1988 precipitate a sharp upturn in demand. It is noticeable that each such 'shock' can cause a lasting period of volatility in a demand line, possibly in the opposite direction as the original shock (as in the top line in Fig. 5). This seems to be because the influence is not significantly dampened but 'rings' around the model and changes become locked-in due to mutually self-reinforcing influence between households. Fig. 6 shows the difference when the network of influence is limited to only adjoining neighbours which cuts out most of the social influence. The pattern is usually regular and predictable for most households –unlike observed patterns of water demand.

It is important that each run can turn out to be different, even though the parameters and set-up is the same. This is due to the fact that, in the Thames Valley in the UK, very similar neighbourhoods (in terms of socio-economic profile and size) can display very different patterns of aggregate water demand. Thus this model is not only intended to capture a *typical* water demand response but the *range* of water demand responses. For this reason one run (or even statistics about many runs) is insufficient to characterise the output from a particular model set-up. Thus in the examples below we will show a set of runs, to give some idea about the range of behaviours that each set-up can exhibit.

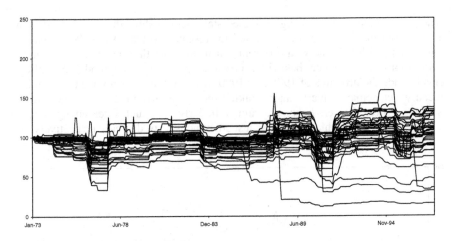

Fig. 6. As Fig. 5 but with neighbourhoods of size 4

Next we change the topology of the social influence network so that the social relations are wrapped-around as if they lived on a torus – this is shown in Fig. 7. This has the effect of increasing the short-term volatility of many of the runs but decreasing the longer-term variation in a run. It seems that the whole population can 'flip' from one behaviour to another – they are too connected. This contrasts with the runs shown in Fig. 5 where different behavioural 'regimes' may be found on different edges of the population. The shape of the neighbourhood also seems to make a difference.

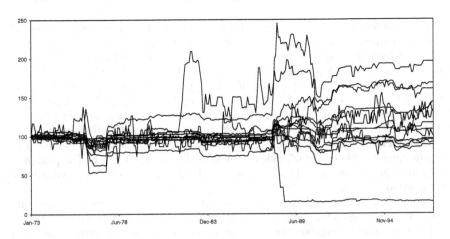

Fig. 7. As Fig. 4 but with wrapped neighbourhoods (rather than unwrapped)

Fig. 8 shows the what happens when the shape of the neighbourhood is changed to a block shape (see Fig. 4). This appears to lessen the impact of droughts.

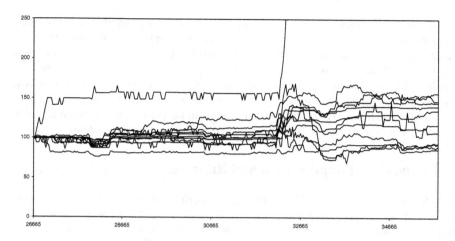

Fig. 8. As Fig. 5 but with block-shaped neighbourhoods (rather than cross-shaped)

Lastly we experimented with the memory coefficient. This halved the rate at which past behaviours are forgotten, this is shown in Fig. 9.

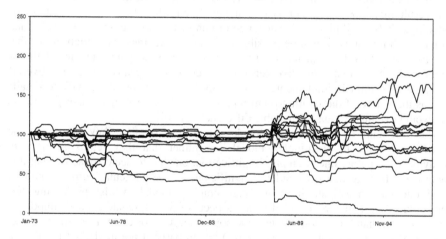

Fig. 9. As Fig. 5 but with a memory coefficient of 5 (rather than 2.5)

Unsurprisingly many of the lines are a lot flatter. In particular many more of the demand patterns reverted after droughts to the levels they were before them. The introduction of innovations (power showers and water-saving washing machines) still has an affect on the general levels.

Elsewhere [3] it was shown that the: climatological data: the timing of innovation introduction; and the proportion of different biases as to influence sources can all make a marked qualitative difference to the aggregate demand patterns that result. However [3] also showed that the *shape* of the initial distribution of the use of appliances in households turned out not to be important and hence could be simplified. Thus this element is a candidate for simplification.

To summarise these results and those from [3]. it appears that the following elements of the model *are* important to the kinds of results that one gets out of the model: the number and size of droughts, the distribution of biases of the households, the timing of innovations, the rate of forgetting, the neighbourhood shape, the topology of the influence grid, and the size of the neighbourhoods. This makes is rather unlikely that a simpler model, that eliminated these would have been descriptively adequate. However it does not rule out the possibility that there are other aspects that might turn out to be unnecessary to produce the desired outputs.

6   Discussion – Simplification and Relevance

MABS models allow for a finer-grained comparison than traditional 'black-box' models. If the outputs from two 'black-box' models are the same given identical inputs, then they are functionally equivalent. In contrast MABS models allow for comparison at finer-grains both in time and in detail. Thus, for example, if two models differ significantly in the behaviour of individual agents *as* they interact then they are different, as with different interactive processes that generate the same aggregate outcomes.

This means that it is extremely unlikely that *any* simplification of a MABS model will result in a *completely* equivalent model. Even when the results appear to be indistinguishable within a particular range of parameters and set-ups, it is likely that, with enough ingenuity, it will be possible to find some settings and initialisations that will force any two different models to diverge in terms of behaviour, and if this is then run for long enough this divergence will become statistically significant. Even if the *same* algorithm is implemented on two different systems, there will be details such as the nature of the floating-point representation and random number generators that would eventually cause detectable differences [1, 14].

Thus the aim of simplifying models should not be that they retain *all* the behaviours of the more complex model, but rather that they retain all *relevant* behaviours. So before one attempts simplification, one has to decide exactly what it is about the behaviour that is considered significant, for the particular purpose one has. Then one can investigate how one can simplify models whilst preserving *this* behaviour. That is, having decided which aspects of the aggregate data are important for the task in hand, one could determine some measures or criteria to test for these. One could then investigate (as I did above) which parts of the model set-up did not effect *these measures or criteria*. One could then simplify away these parts of the model *for this purpose*. In other words, simplification might results from explicit decisions as to representational relevance.

One corollary of this is that it may well be that *different* simplifications will be appropriate when the *same* model is used for *different purposes* or in different contexts. This context-dependency of simplification is opposite to that in the *KISS* approach, for there one makes a model *more complex* in order to be *less general*.

7   Conclusion

The difficult part in science is not finding attractive abstract models, but of relating abstract models to the world (i.e. the target domain). The *KISS* approach ensures that one has an attractive and understandable model, but does not (of itself) give any reason to suppose that it will lead to models that relate strongly enough to the target domain so as to usefully inform us about that domain. The *KIDS* approach starts with a model which relates as strongly to the target domain as possible, but does not ensure that the models are "elegant". Before the advent of cheap computational power, it was only possible to get any results out of analytic (and hence relatively simple models), this made the *KIDS* approach infeasible.

The trade-off between the practicality of our models and their descriptive adequacy is a complex and context-dependent one – as with all modelling decisions, there is no final and general answer [9]. Neither the *KISS* nor the *KIDS* approach will *always* be the best one, and complex mixtures of the two will be frequently appropriate. However the balance is shifting away from *KISS* and towards *KIDS* in areas dominated by complex phenomena. In such areas there is no reason to suppose that elegant models will be particularly useful and the advent of MABS facilitates the creation, management and communication of complex, descriptive models.

In short, when modelling multi-agent and multi-actor systems, (where there are many good reasons to suspect that things will be very complex), one would need strong reasons for adopting a *KISS* methodology – much better reasons than empty invocations of "simplicity". In science, at least, truth comes before beauty.

Acknowledgements

We would like to thank the many people with whom we have discussed these issues, including: Tom Downing, Juliette Rouchier, Jim Doran, David Hales, Rosaria Conte and Guillaume Deffuant. We would also like to thank the anonymous referees whose queries have provided us with the perfect excuse for citing so much of our own work.

References

1. Axtell, R., & al. (1996), Aligning Simulation Models: A Case Study and Results, *Computational and Mathematical Organization Theory* 1:123-141.
2. Barreteau, O. & al. (2001). Role-playing games for opening the black box of multi-agent systems: method and lessons of its application to Senegal River Valley irrigated systems. *Journal of Artificial Societies and Social Simulation* 4(2), <http://jasss.soc.surrey.ac.uk/4/2/5.html>
3. Barthelemy, O. (2003) The impact of the model structure in social simulations, 1st International Conference of the European Social Simulation Association, Gronigen, the Netherlands, September 2003. <http://cfpm.org/cpmrep121.html>
4. Chattoe, E. (2002). Building Empirically Plausible Multi-Agent Systems: A Case Study of Innovation Diffusion. In K. Dautenhahn, & al. (eds.). *Socially Intelligent Agents - creating relationships with computers and robots*. Dordrecht, Kluwer.
5. Downing, T.E, & al. (2003). Climate Change and the Demand for Water, Research Report, Stockholm Environment Institute Oxford Office. <http://www.sei.se/oxford/ccdew>
6. Edmonds, B. (1999). Syntactic Measures of Complexity. Doctoral Thesis, University of Manchester, Manchester, UK. <http://bruce.edmonds.name/thesis>
7. Edmonds, B. (2000). Complexity and Scientific Modelling. *Foundations of Science*, 5:379-390.
8. Edmonds, B. (2000) The Purpose and Place of Formal Systems in the Development of Science, CPM Report 00-75. <http://cfpm.org/cpmrep75.html>

9. Edmonds, B. (2001) The Use of Models - making MABS actually work. In. Moss, S. & Davidsson, P. (eds.), Multi Agent Based Simulation, *Lecture Notes in Artificial Intelligence*, 1979:15-32.
10. Edmonds, B. (2002) Simplicity is Not Truth-Indicative. CPM Report 02-99. <http://cfpm.org/cpmrep99.html>
11. Edmonds, B. (in press) Simulation and Complexity - how they can relate. In Feldmann, V. and Mühlfeld, K. (eds.) *Virtual Worlds of Precision*. Lit Verlag. <http://cfpm.org/cpmrep118.html>
12. Edmonds, B. & Bryson, J. (2004) The Insufficiency of Formal Design Methods - the necessity of an experimental approach for the understanding and control of complex MAS. In Jennings, N. R. et al. (eds.) *Proceedings of the 3rd International Joint Conference on Autonomous Agents & Multi Agent Systems* (AAMAS'04), July 19-23, New York, ACM Press, 938-945.
13. Edmonds, B. & Hales, D. (2003) Computational Simulation as Theoretical Experiment, CPM report 03-106. (submitted to the *Journal of Mathematical Sociology*). <http://cfpm.org/cpmrep106.html>
14. Edmonds, B. and Hales, D. (2003) Replication, Replication and Replication - some hard lessons from model alignment. *Journal of Artificial Societies and Social Simulation* 6(4) <http://jasss.soc.surrey.ac.uk/6/4/11.html>
15. Etienne, M., Le Page, C. & Cohen, M. (2003) A Step-by-step Approach to Building Land Management Scenarios Based on Multiple Viewpoints on Multi-agent System Simulations. *Journal of Artificial Societies and Social Simulation* 6(2) <http://jasss.soc.surrey.ac.uk/6/2/2.html>
16. Giere, R. N. (1988). Explaining Science: A Cognitive Approach. Chicago, University of Chicago Press.
17. Granovetter, M. (1985). Economic-Action and Social-Structure – The Problem of Embeddedness. *American Journal of Sociology* 91:481-510.
18. Moss, S. (1998). Critical Incident Management: An Empirically Derived Computational Model. Journal of Artificial Societies and Social Simulation 1(4), <http://jasss.soc.surrey.ac.uk/1/4/1.html>
19. Moss, S. (2002). Policy Analysis from First Principles. *Proceedings of the US National Academy of Sciences* 99:7267-7274.
20. Moss, S. & Edmonds, B. (accepted) Sociology and Simulation: Statistical and Qualitative Cross-Validation. *American Journal of Sociology*. Earlier version is CPM report 03-105. <http://cfpm.org/cpmrep105.html>
21. Pohill, J. G., Izquierdo, L. R. & Gotts, N. M. (2003) The Ghost in the Model (and other effects of floating point arithmetic). 1st International Conference of the European Social Simulation Association, Gronigen, the Netherlands, September 2003. <http://www.uni-koblenz.de/~kgt/ESSA/ESSA1/Polhill-Izquierdo-Gotts.pdf>
22. Rouchier, J. (2004) Interaction Routines and Selfish Behaviours in an Artificial Market – Transferring field observations of a wholesale fruits and vegetables market into a multi-agent model. CPM Report No.: CPM-04-130. <http://cfpm.org/cpmrep130.html>
23. Suppes, P. (1962). Models of Data. Logic In E. Nagel, P. Suppes & A. Tarski (eds.) *Methodology and the Philosophy of Science: Proceedings of the 1960 International Congress*. Palo Alto, CA, Stanford University Press: 252-261.
24. Taylor, R. (2003) Agent-Based Modelling Incorporating Qualitative and Quantitative Methods: A Case Study Investigating the Impact of E-commerce upon the Value Chain. 1st International Conference of the European Social Simulation Association, Gronigen, the Netherlands, September 2003. <http://cfpm.org/cpmrep123.html>

Analysis of Learning Types in an Artificial Market

Kiyoshi Izumi, Tomohisa Yamashita, and Koichi Kurumatani

ITRI, AIST & CREST, JST, 2-41-6 Aomi, Koto-ku, Tokyo 135-0064, Japan

Abstract. In this paper, we examined the conditions under which evolutionary algorithms (EAs) are appropriate for artificial market models. We constructed three types of agents, which are different in efficiency and accuracy of learning. They were compared using acquired payoff in a minority game, a simplified model of a financial market. As a result, when the dynamics of the financial price was complex to some degree, an EA-like learning type was appropriate for the modeling of financial markets.

1   Introduction

An artificial market approach, a new agent-based approach to financial market research, has been developing against the background of drastic changes of financial markets in recent years. An artificial market study found mechanisms of several market phenomena, such as a financial bubble, which were not able to be explained well by the conventional researches [1, 2]. An artificial market study builds a multi-agent model of a financial market, where an agent trades as a virtual dealer. In many artificial market models, each agent can change its behavior rules by learning based on past performance. Many studies used evolutionary algorithms (EAs), such as Genetic Algorithm (GA) or Genetic Programming (GP), as agents' learning method.

Although artificial market models using EA have achieved many successful results, there is little research which examined whether EA would be appropriate as the learning method in an artificial market model. We analyzed field data such as interviews and questionnaires to actual dealers, and verified appropriateness of EA as a learning method in an artificial market from the empirical viewpoint [3, 1]. Chen and Yeh implemented EA mechanism in their artificial market model, and found that their model had similar characteristics to the actual financial markets such as a fat-tail property and nonlinearity [2]. That is, they showed EA's ability to simulate the actual financial markets.

The purpose of this paper is to verify the appropriateness of applying EA to an artificial market model from a different viewpoint. We focus on the function of a learning algorithm. This paper explores these questions focusing on the relationship between the past financial price movement and the supply and demand. So, this paper conducts computer experiments under a game-theoretic environment called a minority game, a simplified model of a financial market from the viewpoint of this relationship. As learning algorithms of agents who participate in the game, we prepared several kinds of learning types. These learning types differ in input information, accuracy, and efficiency. And we compared the performance of agents that have different learning types, under various conditions. Then it was investigated what conditions made a learning algorithm like EA appropriate for an artificial market.

P. Davidsson et al. (Eds.): MABS 2004, LNAI 3415, pp. 145–158, 2005.

2   Framework of Experiments

2.1   Minority Game as a Model of Financial Markets

A minority game is a repetition game in which N (odd number) players must choose one of two alternatives at each step. A payoff is given to a minority group that consists of players who chose the alternative which fewer people chose between two alternatives. Arthur proposed the idea of a bar problem that people try to drink at the bar which fewer person chose between two bars [4]. Many other researchers had discussed the nonlinearity of this phenomenon and made various extensions [5].

Since the mechanism that a minority group wins is seen also in an actual financial market, a minority game can be considered as a simplified model of a financial market [6]. In this paper, a standard minority game [7,8] was extended as an artificial market model. N (odd number) agents participate in a game, and time progresses dispersedly. One period consists of four steps; (1) Determination of action, (2) Price determination, (3) Calculation of payoffs, and (4) Learning.

(1) Determination of Action: Each agent i determines its dealing action $h^i(t)$, to buy or sell a certain financial capital at t, using knowledge called a *memory* $\mathbf{P}^m(t-1)$. The memory $\mathbf{P}^m(t-1)$ is a time series data of price changes of the financial price at the past m time steps.

$$\mathbf{P}^m(t-1) = \{P(t-1), P(t-2), \cdots, P(t-m)\} \qquad (1)$$

$P(\tau)$ is either $+1$ or -1, and expresses the change in the price at time τ. When the financial price rises or drops at time τ, then $P(\tau)$ equals $+1$ or -1, respectively. Each agent i has the rule that determines its behavior $h^i(t)$ (buy or sell) according to each pattern of the price changes $\mathbf{P}^m(t-1)$. This rule is called strategy $S^i(t)$ (see Fig. 1).

(2) Price Determination: The supply and demand of all N agents are accumulated after all agents determined their action. When more agents try to buy, the price rises. When more agents try to sell, the price falls. That is, the price change $P(t)$ is decided by majority.

$$P(t) = \begin{cases} +1 \text{ up} & (\sum_{i=1}^N h^i(t) > 0) \\ -1 \text{ down} & (\sum_{i=1}^N h^i(t) < 0) \end{cases}$$

This equation simplifies the relation between financial prices and supply and demand in a market.

(3) Calculation of Payoff: A payoff to each agent i, payoff$^i(t)$, is calculated from the last price change $P(t)$ and its dealing behavior $h^i(t)$.

$$\text{payoff}^i(t) = -h^i(t) \cdot \sum_{j=1}^N h^j(t) \qquad (2)$$

By this equation, when the price rises ($P(t) = +1$), agents that sold the financial capital ($h^i(t) = -1$) belong to a minority group, and they acquire a positive payoff. On the other

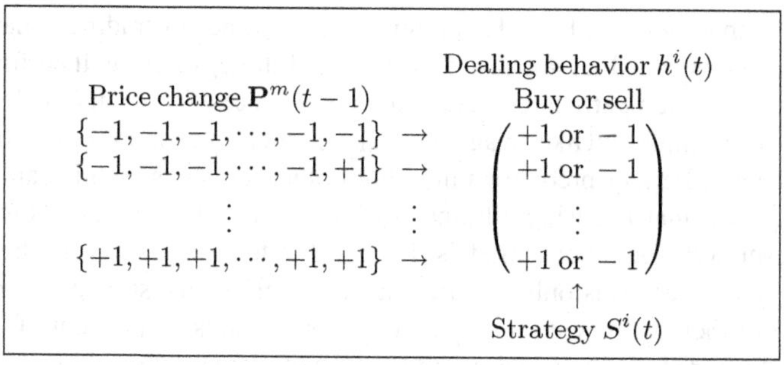

Dealing behavior $h^i(t)$

Price change $\mathbf{P}^m(t-1)$ Buy or sell

$$
\begin{array}{lcl}
\{-1,-1,-1,\cdots,-1,-1\} & \rightarrow & \\
\{-1,-1,-1,\cdots,-1,+1\} & \rightarrow & \\
\vdots & \vdots & \\
\{+1,+1,+1,\cdots,+1,+1\} & \rightarrow &
\end{array}
\left(\begin{array}{c}
+1 \text{ or } -1 \\
+1 \text{ or } -1 \\
\vdots \\
+1 \text{ or } -1
\end{array}\right)
$$

$\uparrow$

Strategy $S^i(t)$

Fig. 1. Each agent's strategy

hand, when the price falls ($P(t) = -1$), agents that bought the capital ($h^i(t) = +1$) belong to a minority group, they acquires a positive payoff.

Economically, the equation 2 assumes that the financial price will returns to the average value in the future. That is, when the price rises (falls), it will fall (rise) to the previous level in the future. And the equation 2 also assumes that the payoff is calculated based on the final value of the financial capital in terms of the price in the future. Therefore, the payoff is positive when a player buys (sells) the financial capital in drop (rise) of the price. The payoff is negative when a player sells (buys) the financial capital in drop (rise) of the price. Under such a regression assumption, a finance market can be considered as a minority game.

(4) Learning: The learning in a standard minority game is simple: selection of strategy. Each player has s strategies generated randomly at the beginning of the game, and continue to have those strategy without modifying them. And each strategy has a specific value called a virtual value. It is a number of times that the dealing behavior derived from a strategy did acquire a positive payoff. Each agent chooses one strategy with the highest value from s strategies, and uses it when the agent determines its behavior at the next step.

In studies of the standard minority game, only the very simple learning was assumed and they analyzed in many cases about the relationship between memory length m and the price fluctuation. However, Cavagna [9] suggested that it has essential significance that all agents are homogeneous, that is, all agents share the same information and the same learning algorithm. He showed that the same results were obtained in both cases that all agents used information about price movement and that all agents used random data. In our study, agents are heterogeneous and the information and the learning methods differ among agents.

2.2 Learning Types

We prepared 4 types of agents; one standard type and three extended types. First, **Chartist (Ch)** was prepared as a standard agent. It is extended from the player in the standard minority game, described in section 2.1. This agent determines its behavior based on the time series of past price changes (chart information).

Besides the standard agent, we prepared three kinds of agent; **Hand imitator (HI)**, **Perfect predictor (PP)**, and **Strategy imitator (SI)**. They are different in terms of kinds of information that they use. First, Hand imitator determines its trading behavior with a value of a parameter, and does not have a complicated strategy or rule. It performs simple learning by copying the parameter value from other successful agents. Its learning and behavior rules are simplified like an agent in many models of econophysics (for example, see [10]). Second, Perfect predictor infers both other agents' strategies and the game structure (a payoff matrix) using all kinds of information. It corresponds to a rational agent in conventional economic models. Finally, Strategy imitator is in the middle of these two types. It performs only an inference of other agents' strategies, and imitates the strategy of other agents with high payoff. It corresponds to an agent of an artificial market model where learning is described by an evolutionary algorithm.

Chartist(Ch). Chartist's behavior decision is the same as stated in section 2.1. It is extended about learning. An agent in the standard minority games continues to have its strategies given first without changing, as described in section2.1. Thus, it can not search for all of solution spaces. Then, we extended Chartist's learning method as follows, to enable it to search for all solution spaces.

Decision of Behavior. Chartist has one strategy described in figure 1. The pattern matching of the price changes of the past m steps, $\mathbf{P}^m(t-1)$, to the strategy $S^i(t)$ determines Chartists' behavior $h^i(t)$.

Learning. When Chartists acquires a positive payoff at t, it does not change its strategy. When it got a negative payoff, its strategy $S^i(t)$ is updated at a certain probability α (a learning rate). That is, when the behavior rule about this price pattern is to buy, the rule is changed to sell, and vice versa.

Hand Imitator (HI). Hand imitator performs simple learning of imitating the behavior which other agents with a high payoff.

Decision of Behavior. According to a certain probability p_{buy}, Hand imitator buys the financial capital. Probability to sell p_{sell} is $1 - p_{buy}$.

Learning. The probability of dealing behavior of other agents with a high payoff is copied.

1. Inference of others' dealing probability
   About other agents js except itself, an estimated probability to buy $\tilde{p}^j_{buy}(t)$ is updated by the following equation.

$$\tilde{p}^j_{buy}(t) = (1 - \beta) \cdot \tilde{p}^j_{buy}(t-1) + \beta \cdot \text{action}^j(t), \tag{3}$$

where $\text{action}^j(t)$ is the agent j's trading behavior at t.

$$\text{action}^j(t) = \begin{cases} 1 & (\text{Agent } j \text{ bought at } t) \\ 0 & (\text{Agent } j \text{ sold at } t) \end{cases}$$

The parameter $0 \leq \beta \leq 1$ expresses the rate which updates the estimated value of the probability of dealing behavior of other agents. β means the learning speed of others' models.

2. Accumulation of payoff

About all agents j including itself, the accumulation value $R^j(t)$ of payoff is updated by the following equation.

$$R^j(t) = (1 - \gamma) \cdot R^j(t-1) + \gamma \cdot \text{payoff}^j(t), \qquad (4)$$

where $\text{payoff}^j(t)$ is the payoff of agent j at t. The parameter $0 \leq \gamma \leq 1$ expresses the update rate of the accumulation value of a payoff, and fixed it to 0.5 in this study.

3. Copy of the behavior according to the payoff

Each agent copies the probability to buy from other successful agents with a certain probability α (a learning rate). First, one agent j^* is chosen by the probability proportional to the accumulation value $R^j(t)$ from all the agents. Next, the estimated trading probability $\tilde{p}^{j^*}_{buy}(t)$ about the agent j^* is copied to its own probability $p^i_{buy}(t)$.

Strategy Imitator (SI). Strategy imitator performs only an estimation of other agents' strategies, and imitates the strategy of other agents with high payoffs.

Decision of Behavior. The dealing behavior $h^i(t)$ is determined by the pattern matching of the price changes of the past m steps, $\mathbf{P}^m(t-1)$, to the strategy $S^i(t)$. It is the same as that of Chartist.

Learning. Strategy imitator estimates others' strategies and imitates the other's strategy with high payoff.

1. Estimation of others' strategies

Strategy imitator estimates whether the other agent js to buy or sell from the pattern matching of the price change $\mathbf{P}^m(t-1)$ to estimated others' strategies $\tilde{S}^j(t)$. If the estimated behavior is different from the actual behavior which agent j performed, by a certain probability $0 \leq \beta \leq 1$ (learning speed of others' models), the estimated strategy $\tilde{S}^j(t)$ will be updated. The bit of the agent j's behavior corresponding to the price change in $\tilde{S}^j(t)$ will be inverted.

2. Accumulation of a payoff

About all agents j including itself, the accumulation value $R^j(t)$ of payoff is updated by the equation 4. It is the same as that of Hand imitator.

3. Imitation of strategy according to the payoff

The other agent's strategy is copied by a certain probability α (learning rate). First, one agent j^* is chosen by the probability proportional to the accumulation value $R^j(t)$ of each payoff from all the agents. Next, the agent's strategy $\tilde{S}^{j^*}(t)$ is copied to its own strategy $S^i(t)$.

Perfect Predictor (PP). Perfect predictor estimates both the other agents' strategies and a game structure (a payoff matrix) using all information.

Decision of Behavior. Perfect predictor estimates others' behavior and decides its own behavior according to the estimated payoff matrix.

1. Estimation of others' behavior
   Perfect predictor estimates the other agent js' behavior (to buy or sell) by the pattern matching of the price changes $\mathbf{P}^m(t-1)$ to estimated the others' strategies $\tilde{S}^j(t)$.
2. Decision of behavior
   Perfect predictor has its own strategy that represents an estimated game structure (a payoff matrix). The strategy means which behavior (buy or sell) can acquire a positive payoff corresponding to the others' behavior. According to the strategy and the other agent j's estimated behavior, Perfect predictor decides its own behavior.

Learning. Perfect predictor learns both others' strategies and the game structure.

1. Learning of others' model
   The estimated strategies $\tilde{S}^j(t)$ about the other agent js are updated. Learning method is the same as that of Strategy imitator.
2. Learning of the game structure
   When each agent acquired a negative payoff, the bit of behavior corresponding to the estimated others' behavior is reversed by a certain probability α (learning rate). This means renewal of the knowledge about the payoff structure.

3   Experiment: Match Against Standard Type

We measured how much better HI, SI, and PP could perform than Chartist, and compared them in performance against Chartist. Each agent type (HI, SI, or PP) plays minority games against the standard agent type, Chartist.

3.1   Setting of the Simulation

The setting of the computer experiments in this paper is shown in Table 1. 25 agents participate in each trial of the minority game. 20 agents are Chartists as standard agents in all trials. The remaining 5 agents are Hand imitators, Strategy imitators, or Perfect predictors.

Table 1. Setting of the simulation

Number of agents	$N = 25$
Agents' combination	{20 CHs and 5 HIs}, {20 CHs and 5 SIs}, or {20 CHs and 5 PPs}
Memory length	$m = \{1, 2, \cdots, 15\}$
Learning speed	$\alpha = 0.8$ (fixed)
Learning speed of others' model	$\beta = \{0.1, 0.2, \cdots, 0.9\}$
Update rate of payoff	$\gamma = 0.5$ (fixed)
Simulation step	1000 steps
Number of simulation runs	10 times every parameter combination
Comparison method	The improvement rate of the average payoff of 5 HIs, 5 SIs, or 5 PPs from the average payoff of 20 CHs.

(Ch: Chartist, HI : Hand imitator, SI: Strategy imitator, and PP : Perfect predictor)

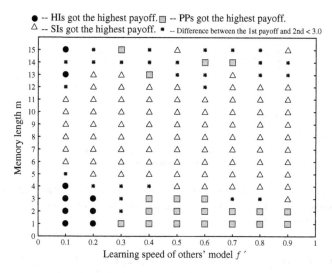

Fig. 2. Summary of results: Each symbol represents which agent type got the highest improvement rate. There are 4 distinct areas ($m \leq 3$ & $\beta < 0.3$, $m \leq 3$ & $\beta > 0.3$, $3 < m < 12$, and $m \geq 13$) according to the winners' types

The learning speed α of all agents' strategy and the update rate of a payoff γ were fixed. The memory length m and the learning speed of others' models β of HI, SI, and PP were changed as shown in Table 1.

The simulation was performed 10 times every parameter set; {Agents' combination (CH vs. HI, SI, or PP) × Memory length m × Learning speed of others' models β}. The agents (HI, SI, and PP) were compared by the improvement rate, $ImpRate$, of the average payoff of 5 HIs, 5 SIs, or 5 PPs from the average payoff of the 20 standard agents (Chartists).

$$
\begin{aligned}
ImpRate(type) &= \frac{\overline{payoff}_{type} - \overline{payoff}_{CH}}{|\overline{payoff}_{CH}|} \times 100, \\
&= \frac{\sum_{i=1}^{5} \text{payoff}_{type}^{i}/5 - \sum_{j=1}^{20} \text{payoff}_{CH}^{j}/20}{|\sum_{k=1}^{20} \text{payoff}_{CH}^{k}/20|} \times 100, \quad (5)
\end{aligned}
$$

where $type$ is an agent type (HI, SI, or PP), $\overline{payoff}_{type}$ is the averaged payoff of $type$, $\overline{payoff}_{CH}$ is the averaged payoff of Chartist, payoff_{type}^{i} is an accumulated payoff of the i-th agent in $type$ at time 1000, and payoff_{CH}^{j} is that of the j-th Chartist. The improvement rates are averaged over 10 simulation runs.

3.2 Simulation Results

The simulation results are summarized in Fig. 2. Different types of learning methods got the highest improvement rate according to the memory length m and the learning speed of others' models β.

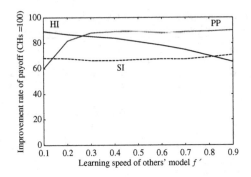

Fig. 3. Simulation results when the memory length m is 2. When the learning speed is slow, the payoff of HI (Hand imitator) is high. When the learning speed of an others model is fast, the payoff of PP (Perfect predictor) is high

Joshi's paper [11] and our preceding study [12] revealed that the memory length linked to the complexity of dynamics of the whole system. When memory length is short, the whole system showed relatively simple patterns that can be described by dynamic systems of finite dimensions. As the memory length got longer, the dynamics patterns became more complex, and finally they can not be described by any finite dimensional dynamic systems[1]. Therefore the memory length m can be considered as an index of complexity of the whole system.

The learning speed of others' models β is an index of the time restriction to learning. All agents continuously change their own strategies at the fixed learning speed α. When β is small, agents must track the others' strategies at a low learning speed, so the time restriction to learning is strong. When β is large, agents can have relatively enough time to track the others' strategies, so the time restriction to learning is weak.

Fig. 2 shows that there are 4 distinct areas according to the two conditions, the complexity and time restriction.

Short Memory Length and Slow Learning Speed Area. When the memory length is short ($m \leq 3$) and the learning speed of others' models is slow ($\beta < 0.3$), Hand imitators' payoff was higher. This result was also shown in Fig. 3. As you can see from the typical results of agents' payoffs in Table 2, Hand imitators could perform against the standard agent the best. The other simulation runs showed similar results. This is because the Perfect predictors and Strategy imitators can not have enough time to learn the others' strategies because of the strong time restriction, and they showed the poor performance[2]. Only Hand imitators could catch up with the fast change of Chartists' strategies

---

[1] LeBaron [13] found that an artificial market with long time horizon agents showed simpler dynamics, in opposition to [11, 12]. That is because he changed the time horizon to evaluate the performance of agents' rules and didn't change the memory length of rules.

[2] In general, learning methods asking for accurate solutions do not always show poor performance even under strong time pressure. Chen and Liao constructed an artificial model considering trading volume and showed various types of relation between the complexity at the micro level and that at the macro level [14].

Table 2. Payoffs of agents in a typical simulation run when $m = 2$ and $\beta = 0.2$. Payoffs in each table are ranked in descending order

(a) 5 HIs and 20 Chs	(b) 5 SIs and 20 Chs	(c) 5 PPs and 20 Chs
$ImpRate(HI) = 86.09$	$ImpRate(SI) = 64.73$	$ImpRate(PP) = 79.09$
$payoff_{HI} = -1348$	$payoff_{SI} = -3464$	$payoff_{PP} = -2054.8$
$payoff_{CH} = -9689.4$	$payoff_{CH} = -9822.4$	$payoff_{CH} = -9828.7$

	Payoff	Type		Payoff	Type		Payoff	Type
1	-912	HI	1	-3148	SI	1	-1502	PP
2	-1110	HI	2	-3268	SI	2	-1870	PP
3	-1316	HI	3	-3340	SI	3	-1904	PP
4	-1690	HI	4	-3410	SI	4	-2408	PP
5	-1712	HI	5	-4154	SI	5	-2590	PP
6	-9220	CH	6	-9388	CH	6	-9298	CH
7	-9372	CH	7	-9402	CH	7	-9418	CH
8	-9398	CH	8	-9428	CH	8	-9506	CH
9	-9400	CH	9	-9432	CH	9	-9522	CH
10	-9432	CH	10	-9544	CH	10	-9586	CH
11	-9464	CH	11	-9686	CH	11	-9590	CH
12	-9650	CH	12	-9704	CH	12	-9730	CH
13	-9670	CH	13	-9740	CH	13	-9766	CH
14	-9678	CH	14	-9750	CH	14	-9798	CH
15	-9708	CH	15	-9756	CH	15	-9816	CH
16	-9734	CH	16	-9900	CH	16	-9898	CH
17	-9736	CH	17	-9910	CH	17	-9906	CH
18	-9846	CH	18	-9926	CH	18	-9934	CH
19	-9848	CH	19	-9944	CH	19	-9972	CH
20	-9852	CH	20	-9960	CH	20	-10036	CH
21	-9876	CH	21	-9984	CH	21	-10038	CH
22	-9906	CH	22	-10076	CH	22	-10120	CH
23	-9932	CH	23	-10118	CH	23	-10142	CH
24	-10026	CH	24	-10178	CH	24	-10190	CH
25	-10040	CH	25	-10622	CH	25	-10308	CH

because of their simple and quick learning method. Moreover, Hand imitators could learn adequately accurate models because the whole system is comparatively simple.

Short Memory Length and Fast Learning Speed Area. When the memory length is short ($m \leq 3$) and the learning speed of others' models is fast ($\beta > 0.3$), Perfect predictor's payoff was high (Fig. 3). Table 3 also shows that Perfect predictors could perform against the standard agent the best under this condition. It is because the Perfect predictors can have enough time to learn the others' strategies and the game structure. Moreover the dynamics of whole systems is relatively simple in this area. They then could obtain exact learning results using all information and beat the other two types, the Hand imitators and Strategy imitators.

Table 3. Payoffs of agents in a typical simulation run when $m = 2$ and $\beta = 0.8$. Payoffs in each table are ranked in descending order

(a) 5 HIs and 20 Chs	(b) 5 SIs and 20 Chs	(c) 5 PPs and 20 Chs
$ImpRate(HI) = 71.57$	$ImpRate(SI) = 69.66$	$ImpRate(PP) = 92.11$
$payoff_{HI} = -2766.4$	$payoff_{SI} = -2855.6$	$payoff_{PP} = -741.2$
$payoff_{CH} = -9729.6$	$payoff_{CH} = -9412.1$	$payoff_{CH} = -9395.1$

	Payoff	Type		Payoff	Type		Payoff	Type
1	-2530	HI	1	-2518	SI	1	-440	PP
2	-2642	HI	2	-2644	SI	2	-730	PP
3	-2710	HI	3	-2870	SI	3	-742	PP
4	-2940	HI	4	-2880	SI	4	-810	PP
5	-3010	HI	5	-3366	SI	5	-984	PP
6	-9096	CH	6	-8772	CH	6	-8944	CH
7	-9406	CH	7	-9026	CH	7	-8974	CH
8	-9512	CH	8	-9058	CH	8	-9064	CH
9	-9514	CH	9	-9130	CH	9	-9150	CH
10	-9570	CH	10	-9270	CH	10	-9158	CH
11	-9600	CH	11	-9338	CH	11	-9158	CH
12	-9612	CH	12	-9376	CH	12	-9240	CH
13	-9612	CH	13	-9388	CH	13	-9258	CH
14	-9614	CH	14	-9390	CH	14	-9302	CH
15	-9618	CH	15	-9462	CH	15	-9324	CH
16	-9666	CH	16	-9486	CH	16	-9376	CH
17	-9672	CH	17	-9510	CH	17	-9386	CH
18	-9682	CH	18	-9522	CH	18	-9490	CH
19	-9832	CH	19	-9528	CH	19	-9498	CH
20	-9860	CH	20	-9580	CH	20	-9662	CH
21	-9874	CH	21	-9600	CH	21	-9728	CH
22	-10012	CH	22	-9602	CH	22	-9736	CH
23	-10148	CH	23	-9634	CH	23	-9810	CH
24	-10250	CH	24	-9738	CH	24	-9822	CH
25	-10442	CH	25	-9832	CH	25	-9822	CH

Medium Memory Length Area. The Strategy imitators' payoff becomes high as memory length becomes longer ($3 < m < 13$). Fig. 4 and Table 4 shows that the Strategy imitators could beat the other two type throughout all learning speed β. When other agents' strategy is complicated, simple learning methods such as the Hand imitators can only acquire inaccurate learning results. And learning methods using all information such as the Perfect predictors spend too much time to get learning results at a limited learning speed. Therefore, payoff of Hand imitator and Perfect predictor had fallen in this area. Then, the Strategy imitators, that are in the middle of these two learning types, got high payoff in this area.

Long Memory Length Area. In this area, the dynamic structure of the whole system is very complicated and then can not be described by any finite dimensional dynamic systems. That is, the system showed chaotic features. Therefore, the difference of per-

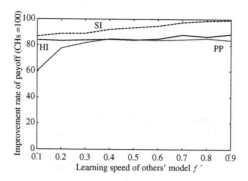

Fig. 4. Simulation results when the memory length m is 9. The payoff of SI (Strategy imitator) becomes high as memory length becomes long

Table 4. Payoffs of agents in a typical simulation run when $m = 9$ and $\beta = 0.5$. Payoffs in each table are ranked in descending order

(a) 5 HIs and 20 Chs	(b) 5 SIs and 20 Chs	(c) 5 PPs and 20 Chs
$ImpRate(HI) = 85.35$	$ImpRate(SI) = 94.7$	$ImpRate(PP) = 84.18$
$\overline{payoff}_{HI} = -947.6$	$\overline{payoff}_{SI} = -326.4$	$\overline{payoff}_{PP} = -981.6$
$\overline{payoff}_{CH} = -6469.1$	$\overline{payoff}_{CH} = -6164$	$\overline{payoff}_{CH} = -6206.2$

	Payoff	Type		Payoff	Type		Payoff	Type
1	-528	HI	1	222	SI	1	-338	PP
2	-722	HI	2	-232	SI	2	-670	PP
3	-1082	HI	3	-352	SI	3	-1256	PP
4	-1142	HI	4	-476	SI	4	-1290	PP
5	-1264	HI	5	-794	SI	5	-1354	PP
6	-6250	CH	6	-5848	CH	6	-5864	CH
7	-6252	CH	7	-5854	CH	7	-5874	CH
8	-6266	CH	8	-5900	CH	8	-5994	CH
9	-6300	CH	9	-5920	CH	9	-6026	CH
10	-6302	CH	10	-5948	CH	10	-6048	CH
11	-6362	CH	11	-5996	CH	11	-6072	CH
12	-6370	CH	12	-6006	CH	12	-6124	CH
13	-6402	CH	13	-6042	CH	13	-6206	CH
14	-6432	CH	14	-6146	CH	14	-6246	CH
15	-6468	CH	15	-6184	CH	15	-6248	CH
16	-6484	CH	16	-6186	CH	16	-6280	CH
17	-6486	CH	17	-6224	CH	17	-6300	CH
18	-6488	CH	18	-6236	CH	18	-6310	CH
19	-6602	CH	19	-6262	CH	19	-6324	CH
20	-6610	CH	20	-6280	CH	20	-6328	CH
21	-6622	CH	21	-6290	CH	21	-6334	CH
22	-6630	CH	22	-6306	CH	22	-6362	CH
23	-6646	CH	23	-6324	CH	23	-6370	CH
24	-6692	CH	24	-6632	CH	24	-6404	CH
25	-6718	CH	25	-6696	CH	25	-6410	CH

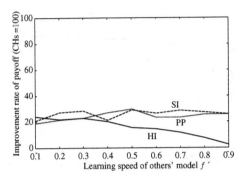

Fig. 5. Simulation results when the memory length m is 13. The performance of all agent types have small difference

Table 5. Payoffs of agents in a typical simulation run when $m = 13$ and $\beta = 0.5$. Payoffs in each table are ranked in descending order

(a) 5 HIs and 20 Chs	(b) 5 SIs and 20 Chs	(c) 5 PPs and 20 Chs
$ImpRate(HI) = 30.7$	$ImpRate(SI) = 33.84$	$ImpRate(PP) = 29.02$
$\overline{payoff}_{HI} = -1075.2$	$\overline{payoff}_{SI} = -867.2$	$\overline{payoff}_{PP} = -978$
$\overline{payoff}_{CH} = -1551.6$	$\overline{payoff}_{CH} = -1310.8$	$\overline{payoff}_{CH} = -1377.9$

	Payoff	Type		Payoff	Type		Payoff	Type
1	-924	HI	1	-718	SI	1	-826	PP
2	-1066	HI	2	-826	SI	2	-872	PP
3	-1096	HI	3	-860	SI	3	-898	PP
4	-1106	HI	4	-912	SI	4	-1042	CH
5	-1184	HI	5	-1020	SI	5	-1082	PP
6	-1220	CH	6	-1036	CH	6	-1144	CH
7	-1242	CH	7	-1128	CH	7	-1212	PP
8	-1310	CH	8	-1172	CH	8	-1238	CH
9	-1336	CH	9	-1186	CH	9	-1250	CH
10	-1348	CH	10	-1202	CH	10	-1258	CH
11	-1438	CH	11	-1206	CH	11	-1260	CH
12	-1498	CH	12	-1222	CH	12	-1310	CH
13	-1534	CH	13	-1222	CH	13	-1322	CH
14	-1542	CH	14	-1246	CH	14	-1340	CH
15	-1562	CH	15	-1268	CH	15	-1340	CH
16	-1566	CH	16	-1330	CH	16	-1352	CH
17	-1584	CH	17	-1364	CH	17	-1352	CH
18	-1590	CH	18	-1366	CH	18	-1388	CH
19	-1604	CH	19	-1372	CH	19	-1452	CH
20	-1690	CH	20	-1394	CH	20	-1486	CH
21	-1734	CH	21	-1420	CH	21	-1502	CH
22	-1740	CH	22	-1462	CH	22	-1556	CH
23	-1816	CH	23	-1502	CH	23	-1596	CH
24	-1830	CH	24	-1546	CH	24	-1622	CH
25	-1848	CH	25	-1572	CH	25	-1748	CH

formance became small among all the 4 agent types; Pattern matchers, Hand imitators, Strategy imitators, and Perfect predictors. Fig. 5 revealed that the improvement rate against the Chartists became small and it becomes hard to distinguish among all agent types by performance. Table 5 also shows there was little difference among the improvement rates of HI, SI, and PP. Moreover, these types could not win over Chartist by a large margin.

We conducted competitions in mixed populations with all 4 agent types, { 6 HIs, 6 SIs, 6 PPs, and 7 CHs }, in order to test the validity of the 4 areas under different combinations [3]. As a result, the similar results were acquired as above mentioned. In the short memory length and slow learning speed area, Hand imitators could acquire the highest payoff. Perfect predictors could get the highest payoff in the the short memory length and fast learning speed area. In the medium memory length area, Strategy imitators won and in the long memory length area there were little difference among payoffs of all the 4 types.

4   Conclusions

Our simulation results revealed that the 4 areas existed from the viewpoint of the advantageous learning types, in the minority game which is the model of a financial market. Which area is applicable for the modeling of a financial market by an artificial market?

Our preceding paper has suggested that it belongs to the medium memory length area [12]. We performed the artificial market simulation using the agent with various memory length. As a result, while the memory was short, agents with longer memory could predict more accurately and were able to get the higher profit. Therefore, it was advantageous to have a longer memory. As each agent used a longer memory, however, the movement of the financial price in the artificial market became more complicated. Then, the movement of the price became complicated too much and the advantageousness of the longer memory disappeared. Thus, each agent stops to make its memory longer before the dynamics of the market become random.

Therefore, it is thought that the modeling by an artificial market belongs to the medium memory length area. In this area, it is reasonable that an agent's learning type is Strategy imitator. If the estimation of other agents' strategies is sufficiently accurate, Strategy imitator is correspond to an agent in an artificial market model where the learning is described by the evolutionary algorithm. Hence, when the above-mentioned conditions are satisfied, the EA mechanism is appropriate for the description of learning in an artificial market model.

References

1. Izumi, K., Ueda, K.: Phase transition in a foreign exchange market: Analysis based on an artificial market approach. IEEE Transactions on Evolutionary Computation **5** (2001) 456–470
2. Chen, S.H., Yeh, C.H.: Evolving traders and the buisiness school with genetic programming: a new architecture of the agent-based artificial stock market. Journal of Economic Dynamics and Control **25** (2001) 363–393
3. Izumi, K., Nakamura, S., Ueda, K.: Development of an artificial market model based on a field study. Information Science (in press)

4. Arthur, W.B.: Inductive reasoning and bounded rationality (the el farol problem). American Economic Review **84** (1994) 406
5. Minority Game's web page: (http://www.unifr.ch/econophysics/)
6. Zhang, Y.C.: Modeling market mechanism with evolutionary games. Europhys. News **29** (1998) 51–54
7. Challet, D., Zhang, Y.C.: Emergence of cooperation and organization in an evolutionary game. Physica A **246** (1997) 407–418
8. Marsili, M.: Market mechanism and expectations in minority and majority games. Physica A **299** (2001) 93–103
9. Cavagna, A.: Irrelevance of memory in the minority game. PHYSICAL REVIEW E **59** (1999) R3783–R3786
10. Lux, T., Marchesi, M.: Scaling and criticality in a stochastic multi-agent model of a financial market. Nature **397** (1999) 493–500
11. Joshi, S., Parket, J., Bedau, M.A.: Technical trading creates a prisoner's dilemma: Results from an agent-based model. In Abu-Mostafa, Y.S., LeBaron, B., Lo, A.W., Weigend, A.S., eds.: Computational Finance 1999, MIT Press (2000) 465–479
12. Izumi, K.: Complexity of agents and complexity of markets. In Terano, T., Nishida, T., Namatame, A., Tsumoto, S., Osawa, Y., T.Washio, eds.: New Frontiers in Artificial Intelligence. Springer (2001) 110–120
13. LeBaron, B.: Evolution and time horizons in an agent based stock market. Macroeconomic Dynamics **5** (2001) 225–254
14. Chen, S.H., Liao, C.C.: Agent-based computational modeling of the stock price-volume relation. Information Sciences (in press)

Toward Guidelines for Modeling Learning Agents in Multiagent-Based Simulation: Implications from Q-Learning and Sarsa Agents

Keiki Takadama[1,2] and Hironori Fujita[3]

[1] Tokyo Institute of Technology,
4259 Nagatsuta-cho, Midori-ku, Yokohama 226-8502 Japan
keiki@dis.titech.ac.jp
[2] ATR Network Informatics Labs,
2-2-2 Hikaridai, "Keihanna Science City" Kyoto 619-0288 Japan
keiki@atr.jp
[3] Hitotsubashi University,
2-1 Naka, Kunitachi-shi, Tokyo 186-8601 Japan
cm040226@srv.cc.hit-u.ac.jp

Abstract. This paper focuses on how simulation results are *sensitive* to agent modeling in multiagent-based simulation (MABS) and investigates such sensitivity by comparing results where agents have different *learning mechanisms*, *i.e.*, Q-learning and Sarsa, in the context of reinforcement learning. Through an analysis of simulation results in a bargaining game as one of the canonical examples in game theory, the following implications have been revealed: (1) even a slight difference has an essential influence on simulation results; (2) testing in static and dynamic environments highlights the different tendency of results; and (3) three stages in both Q-learning and Sarsa agents (*i.e.*, (a) competition; (b) cooperation; and (c) learning impossible) are found in the dynamic environment, while no stage is found in the static environment. From these three implications, the following very *rough* guidelines for modeling agents can be derived: (1) cross-element validation for specifying key factors that affect simulation results; (2) a comparison of results between the static and dynamic environments for determining candidates to be investigated in detail; and (3) sensitive analysis for specifying applicable range for learning agents.

Keywords: Multiagent-based simulation, sensitivity, agent modeling, learning mechanism, bargaining game.

1 Introduction

Veri cation and validation (V&V) is a critical issue in multiagent-based simulation (MABS) [1, 7] due to the fact that simulation results are very sensitive to how agents are modeled. In particular, this problem becomes serious when an agent has a learning mechanism because it causes a complex interaction among

P. Davidsson et al. (Eds.): MABS 2004, LNAI 3415, pp. 159–172, 2005.

agents that yields emergent phenomena in social simulations. Because of these difficulties, our previous research focused on learning mechanisms applied to agents and compared the results of computational models that employ either of (1) evolutionary strategy (ES) [3], (2) learning classifier system (LCS) [4,6], or (3) reinforcement learning (RL) [18]. Through a comparison of the results in game theory [8,9], our research concluded that agents with some learning mechanisms acquire rational behaviors while those with other learning mechanisms acquire human-like behaviors, even though the agent architecture is the same except for the learning mechanisms [19].

The above research found that different kinds of learning mechanisms (i.e., ES, LCS, and RL) may derive different results, even though they try to maximize their profit. Therefore, the next question is what is the result when employing the same kind of learning mechanisms. An example includes Q-learning [21] and Sarsa [15,17], which are both reinforcement learning and differ very little (the detailed difference is shown in equations (1) and (2) in Section 3). Related to this issue, Sutton and Barto, from the computer science field (in particular, machine learning literature), reported that simulation results of Q-learning differ from those of Sarsa [18]. Specifically, Q-learning agents acquire optimal behaviors paying no attention to risk (e.g., danger of receiving large negative rewards), while Sarsa agents cannot acquire optimal behaviors but acquire behaviors that avoid risk. Since the study investigated the results in a single agent environment (i.e., static environment), this paper aims at conducting simulations in a multi-agent environment (i.e., dynamic environment), as is usual in social simulations, and investigating what kinds of differences are found by comparing both environments. Based on such comparison, this paper explores guidelines for modeling learning agents in multiagent-based simulation.

This paper is organized as follows. Section 2 explains the bargaining game as an example for social simulations and an implementation of agents is described in Section 3. Section 4 presents computer simulations and Section 5 discusses a comparison of the results. Finally, our conclusions are given in Section 6.

2   Bargaining Game

As a concrete domain, we focus on bargaining theory [8,9] and employ a bargaining game [14] where two or more players try to reach a mutually beneficial agreement through negotiations. This game has been proposed for investigating when and what kinds of offers of an individual player can be accepted by the other players. We selected this domain because (1) this game is a canonical example; and (2) since the rational behaviors of players have already been analyzed in game theory [12], we can validate simulation results by comparing the rational behaviors of players. Note that such comparisons only validate a certain aspect of results, because rational behaviors analyzed in game theory are not the same as human behaviors analyzed in experimental economics [11,5, 10,13]. But, this research starts by conducting the validation based on rational behaviors.

To understand the bargaining game, let us give an example from Rubinstein's work [14] which illustrated a typical situation using the following scenario: two players, P_1 and P_2, have to reach an agreement on the partition of a "pie". For this purpose, they alternate offers describing possible divisions of the pie, such as "P_1 receives x and P_2 receives $1 - x$ at time t", where x is any value in the interval $[0, 1]$. When a player receives an offer, the player decides whether to accept it or not. If the player accepts the offer, the negotiation process ends, and each player receives the share of the pie determined by the concluded contract. If the player decides not to accept the offer, on the other hand, the player makes a counter-offer, and all of the above steps are repeated until a solution is reached or the process is aborted for some external reason (e.g., the number of negotiation processes is finite). If the negotiation process is aborted, neither player can receive any share of the pie.

Here, we consider the finite-horizon situation, where the maximum number of steps (MAX_STEP) in the game is fixed and all players know this information as common knowledge [16]. In the case where MAX_STEP $= 1$ (also known as the ultimatum game), player P_1 makes the only offer and P_2 can accept or refuse it. If P_2 refuses the offer, both players receive nothing. Since a rational player is based on the notion of "anything is better than nothing", a rational P_1 tends to keep most of the pie to herself by offering only a minimum share to P_2. Since there are no further steps to be played in the game, a rational P_2 inevitably accepts the tiny offer.

By applying a backward induction reasoning to the situation above, it is possible to perform a simulation for MAX_STEP > 1. For the same reason seen in the ultimatum game, the player who can make the last offer is better positioned to receive the larger share by offering a minimum offer [16]. This is because both players know the maximum number of steps in the game as common knowledge, and therefore the player who can make the last offer can acquire a larger share with the same behavior of the ultimatum game at the last negotiation. The point of multiple-step negotiation is to investigate whether the advantageous player can continue the negotiation to the last step to acquire a larger share under the situation. From this feature of the game, the last offer is granted to the player who does not make the first offer if MAX_STEP is even, since each player is allowed to make at most MAX_STEP/2 offers. On the other hand, the last offer is granted to the same player who makes the first offer if MAX_STEP is odd.

After this section, we use the terms "payoff" and "agent" instead of the terms "share" and "player" for their more general meanings in the bargaining game.

3   Modeling Agents

To implement reinforcement learning agents in the framework of the bargaining game described in the previous section, we implement agents as follows.

- **Strategies memory** stores a fixed number of matrixes of offers (O) and thresholds (T) as shown in Figure 1. In particular, the MAX_STEP/2+1 number

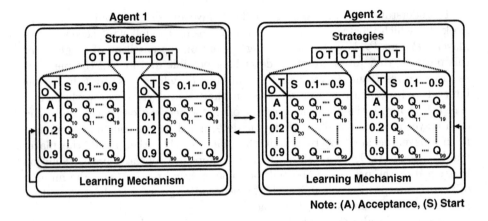

Note: (A) Acceptance, (S) Start

Fig. 1. Reinforcement learning agents

of matrixes are used in turn at each negotiation to decide to accept an offer
or make an counter-offer (see an example presented later in this section).
In this model, agents independently learn and acquire different worths of
strategies (i.e., different Q-values in offer and threshold).[1] Note that (1)
both offer and threshold values are represented by the discrete values in a
0.1 unit; and (2) in addition to these 0.1-0.9 values, the matrix has a column
labelled (S) and a row labelled (A), which is used to determine the value of
the first offer and is used to hold the value of accepting an offer, respectively.

– **Learning mechanism** (i.e., Q-learning and Sarsa) updates the worth of
pairs of offer and threshold by the following conventional equations (1) and
(2), respectively. In these equations, $Q(t, o)$, $Q(t', o')$, r, $O(t')$, $\alpha(0 < \alpha \leq 1)$,
and $\gamma(0 \leq \gamma \leq 1)$ indicate the worth of selecting the offer (o) at threshold
(t), the worth of selecting 1 step next offer (o') at 1 step next threshold (t'),
the reward corresponding to the acquired payoff, a set of possible offers at 1
step next threshold (t'), the learning rate, and the discount rate, respectively.
Note that the only difference between Q-learning and Sarsa is whether the
m ax operation is used or not in the equations of updating Q-values.

$$Q - learning : Q(t, o) = Q(t, o) + \alpha[r + \gamma \max_{o' \in O(t')} Q(t', o') - Q(t, o)] \quad (1)$$

$$Sarsa : Q(t, o) = Q(t, o) + \alpha[r + \gamma Q(t', o') - Q(t, o)] \quad (2)$$

The action selection (acceptance or counter-o er) in this paper is based on
the ϵ-greedy m ethod, which selects an action of the maximum worth (Q-

---

[1] In the context of reinforcement learning, worth is called "value". We select the term
"worth" instead of "value" because the term "value" is used as a numerical number
that represents the offer and threshold in strategies.

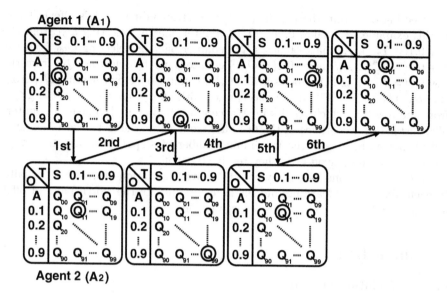

Fig. 2. Example of a negotiation process

value) at the $1 - \epsilon$ probability, while selecting an action randomly at the $\epsilon(0 \leq \epsilon \leq 1)$ probability.

As a concrete negotiation process, agents proceed as follows. Defining $\{O, T\}_i^{A\{1,2\}}$ as the ith offer or threshold value of agent A_1 or A_2, A_1 starts by selecting one Q-value from the row $S(Start)$ (i.e., one Q-value from $\{Q_{10}, \cdots, Q_{90}\}^2$ in the row S), and makes the first offer $O_1^{A_1}$ according to the selected Q-value (for example, A_1 makes an offer 0.1 if it selects Q_{10}). Here, we count one step when either agent makes an offer. Then, A_2 selects one Q-value from the row $T_2^{A_2} (= O_1^{A_1})$ (i.e., one Q-value from $\{Q_{0T}, \cdots, Q_{9T}\}$, where $T = T_2^{A_2} (= O_1^{A_1})$). A_2 accepts the offer if Q_{0T} (i.e., the acceptance (A)) is selected; otherwise, it makes a counter-offer $O_2^{A_2}$ according to the selected Q-value as the same way of A_1. This cycle is continued until either agent accepts the offer of the other agent or a negotiation is over (i.e., the maximum number of steps (MAX_STEP) is exceeded by deciding to make a counter-offer instead of acceptance at the last negotiation step).

To understand this situation, let us consider the simple example where MAX_STEP = 6 as shown in Figure 2. Following this example, A_1 starts to make an offer $0.1(= O_1^{A_1})$ to A_2 by selecting Q_{10} from the row $S(start)$. However, A_2 does not accept the first offer because it determines to make $0.1(= O_2^{A_2})$ counter-offer

---

[2] At the first negotiation, one Q-value is selected from $\{Q_{10}, \cdots, Q_{90}\}$ not from $\{Q_{00}, Q_{10}, \cdots, Q_{90}\}$. This is because the role of the first agent is to make the first offer and not to accpet any offer (by selecting Q_{00}) due to the fact that a negotiation has not started yet.

by selecting Q_{11} from the row $0.1(= T_2^{A_2}$, corresponding to A_1's offer). Then, in this exapmle, A_1 makes $0.9(= O_3^{A_1})$ counter-offer by selecting Q_{91} from the row $0.1(= T_3^{A_1})$, A_2 makes $0.9(= O_4^{A_2})$ counter-offer by selecting Q_{99} from the row $0.9(= T_4^{A_2})$, A_1 makes $0.1(= O_5^{A_1})$ counter-offer by selecting Q_{19} from the row $0.9(= T_5^{A_1})$, and A_2 makes $0.1(= O_6^{A_2})$ counter-offer by selecting Q_{11} from the row $0.9(= T_6^{A_2})$. Finally, A_1 accepts the 6th offer from A_2 by selecting Q_{01} from the row 0.1, which results in A(acceptance). But, if A_1 makes a counter-offer instead of accpetance of the 6th offer from A_2 at the last negotiation step (which means to exceed the maximum number of steps), both agents can no longer receive any payoff, i.e., they receive 0 payoff.

Here, we count one iteration when the above negotiation process ends or fails. In each iteration, Q-learning and Sarsa agents update the worth pairs of offer and threshold in order to acquire a large payoff.

4   Simulation

4.1   Simulation Design

The following two simulations were conducted as comparative simulations.

- **Case 1: Q-learning vs. Sarsa in a static environment**
  Investigation of the results of two different learning mechanisms in a single agent environment, where one agent is a predetermined agent without learning mechanisms. Precisely, we compare results of the predetermined agent vs. Q-learning with those of the predetermined agent vs. Sarsa. The predetermined agent in this case is set as A_1 who makes the first offer, and it is implemented by (1) accepting the offer from A_2 if the offer value is 0.2 or more; otherwise, it makes a 0.1 counter-offer to A_2; and (2) accepting any offer at the last steps (i.e., the maximum number of steps). This implementation is based on rational behaviors analyzed in the bargaining game. In particular, the predetermined agent (A_1) accepts a 0.2 or more offer value from A_2, because the maximum values that A_1 can receive is theoretically calculated as 0.1.
- **Case 2: Q-learning vs. Sarsa in a dynamic environment**
  Investigation of the results of two different learning mechanisms in a multi-agent environment, as is usual in social simulations. Precisely, we compare results of Q-learning vs. Q-learning with those of Sarsa vs. Sarsa.

In each simulation, (a) the payoff and (b) the negotiation process size are investigated by varying the ϵ parameter in the ϵ-greedy method, which determines a randomness of agent behaviors. Here, the negotiation process size is the number of steps until an offer is accepted or MAX_STEP if no offer is accepted. All simulations are conducted for up to 10,000,000 iterations, which is enough for the agents to learn appropriate behaviors, and the results show an average over the last 10,000 iterations, which means the converged values calculated by the average from 9,990,000 to 10,000,000 iterations. We confirmed that both standard deviations of (a) the payoff and (b) the negotiation process size from 9,990,000 to 10,000,000 iterations are almost 0.

As to the parameter setting, the variables are set as follows: MAX_STEP (maximum number of steps in one iteration) is 6; α (learning rate) is 0.1; γ (discount rate) is 1.0; and ϵ (ϵ-greedy method) is from 0.0 to 0.2. Note that preliminary examinations found that the tendency of the results does not drastically change according to the parameter setting except for the ϵ parameter.

All simulations in this paper were implemented in the C language with standard libraries and were conducted using the Linux OS on a Pentium 2.4GHz Processor. Note that the same simulation can be implemented and conducted using other compilers and platforms.

4.2   Simulation Results

Figure 3 shows the simulation results of both the static (case 1) and dynamic (case 2) environments. The upper figures indicate the payoff, while the lower figures indicate the negotiation process size. The vertical axis in the figures indicates these two criteria, while the horizontal axis indicates the ϵ parameter of the ϵ-greedy method. In each figure, the solid line indicates the results of the Q-learning related, while the dotted line indicates those of the Sarsa related. Here, Q-learning related includes two cases of (i) the predetermined agent vs. Q-learning and (ii) Q-learning vs. Q-learning, while Sarsa related includes two cases of (i) the predetermined agent vs. Sarsa and (ii) Sarsa vs. Sarsa. The "Q", "Sarsa", and "Predetermined" in the figures indicate Q-learning, Sarsa, and the predetermined agents, respectively. Furthermore, the "Predetermined-Q", for example, indicates that the predetermined agent corresponds to A_1 while the Q-learning agent corresponds to A_2. Specifically, the payoff of A_1 is shown in the lower lines, while that of A_2 is shown in the upper lines.

These results suggest us that simulation results are affected by both learning mechanisms applied to agents (i.e., Q-learning or Sarsa) and environments (i.e., static and dynamic environments).

5   Discussion

5.1   Q-Learning Versus Sarsa

When comparing the results between Q-learning and Sarsa, Figure 3 shows that (1) the payoff of Q-learning related (i.e., Predetermined-Q or Q-Q) is almost the same as that of Sarsa related (i.e., Predetermined-Sarsa or Sarsa-Sarsa); and (2) the negotiation process size of Q-learning related differs from that of Sarsa related. In particular, we are surprised that the different results occur in the negotiation process size because the difference between Q-learning and Sarsa is very little (i.e., the only difference is whether the m ax operation is used in an equation of updating Q-values as shown in equations (1) and (2)). So, why do we obtain such different results? The reasons for the above results are summarized as follows:

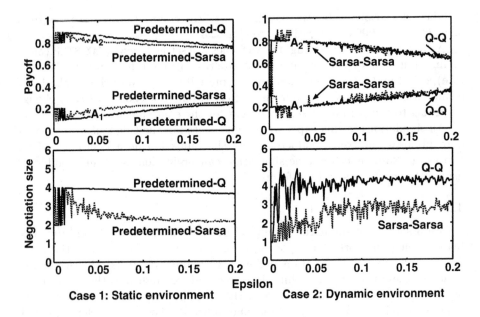

Fig. 3. Simulation results of static and dynamic environments $(0 < \epsilon \leq 0.2)$

- **Payoff:** The payoffs of both Q-learning and Sarsa are mostly the same because agents in the bargaining game are designed not to take large negative
  rewards (i.e., only zero rewards in the worst case) when exceeding the last
  negotiation steps. If agents take large negative rewards, our preliminary simulation found that the payoff of Q-learning becomes small than that of Sarsa.
  This is because Q-learning agents takes risk while Sarsa agents avoid risk,
  which can be understood from implication found by Sutton and Barto in the
  cliff walking problem [18]. It reports that Q-learning agents acquire optimal
  behaviors paying no attention to risk (i.e., danger of receiving large negative
  rewards in the bargaining game), while Sarsa agents cannot acquire optimal
  behaviors but acquire behaviors that avoid risk. Here, the bargaining game
  are designed not to take large negative rewards, and therefore the difference
  in payoffs between Q-learning and Sarsa does not occur.
- **Negotiation process size:** The reason why the negotiation process size
  of Q-learning differs from that of Sarsa can be also understood from the
  same implication that Q-learning takes risk while Sarsa avoid risk. Here,
  considering the risk from the viewpoint of the bargaining game, A_2 has a
  chance to obtain the maximum payoff (i.e., 90% of the rewards) when A_2
  reaches the last negotiation steps, but A_2 should take a risk to do so because
  A_2 cannot receive any share of the rewards if A_1 refuses the offer from A_2.
  In this situation, the negotiation process size becomes large if an agent takes
  risk like Q-learning, while the size becomes small if an agent avoids risk like
  Sarsa. This indicates that Q-learning and Sarsa have different capabilities
  from the viewpoint of risk. Furthermore, what should be noticed here is that

this implication is found not only in the static environment, which Sutton and Barto addressed, but also in the dynamic environment.

We can first conclude that even a slight difference (i.e., whether the m ax operation is used or not in an equation of updating Q-values) has an essential influence on simulation results.

5.2   Static Versus Dynamic Environments: Short Range of ϵ

When comparing the results between the static and dynamic environments, Figure 3 shows that (1) the payoffs of A_1 (or A_2) in both the static and dynamic environments increases (or decreases) as the ϵ parameter increases; and (2) the negotiation process size in the static environment decreases, while that in the dynamic environment increases. Why do we obtain such different results? The reasons for the above results are summarized as follows:

- **Payoff:** The payoffs in both the static and dynamic environments are very similar for the same reason described in Section 5.1. Specifically, the payoff of A_1 (or A_2) increases (or decreases) as the ϵ parameter increases because agents have difficulty in performing rational behaviors due to an increase of the probability of random behaviors.
- **Negotiation process size:** The different results are obtained in the negotiation process size, because the probability of reaching the last negotiation steps in the static and dynamic environments are different. Precisely, Q-learning and Sarsa agents become to make the large offer by mistake instead of making the tiny offer as the ϵ parameter increases. In the static environment, the predetermined agent perfectly accepts such offer even before reaching the last negotiation steps. This makes the negotiation process size decrease. In the dynamic environment, on the other hand, both learning agents have the chance to acquire the large offer when one of agents makes the large mistaken offer. This leads agents to wait for the large mistaken offer from the opponent agent, which makes the negotiation process step increase.

We can secondly conclude that the static and dynamic environments derive the different tendency in the negotiation process size but not in the payoff. This tendency is common in both Q-learning and Sarsa agents.

5.3   Static Versus Dynamic Environments: Long Range of ϵ

In order to investigate the different tendency of the static and dynamic environments in more detail, we extend the simulations by changing the ϵ parameter from 0 to 1 (the ϵ parameter in Figure 3 is set from 0 to 0.2). Figure 4 shows the extended simulation results, where the vertical axis indicates the negotiation process size, while the horizontal axis has the same meaning as in Figure 3. Note that the results in Figure 4 are smoothed in comparison with those in Figure 3 in order to make clear the tendency of results.

These results suggest that (1) no stage is found in the static environment, while (2) three stages (a), (b), and (c) are found in the dynamic environment.

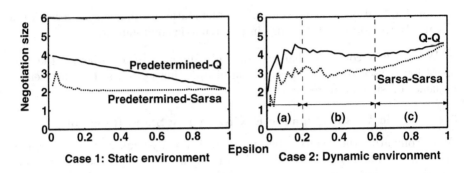

Fig. 4. Simulation results of static and dynamic environments ($0 < \epsilon \leq 1$)

Precisely, the negotiation process size in the static environment mainly decreases as the ϵ parameter increases, while the size in the dynamic environment increases in stage (a), decreases or keeps in stage (b), and increases again in stage (c). These three stages are categorized as follows:

- **(a) Competition stage:** As described in the previous section, agents in this stage have the chance to acquire the large offer when one of agents makes the large mistaken offer. This makes agents com pete with each other to acquire the large mistaken offer by waiting for it from the opponent agent until the last negotiation steps. Since the probability of a mistake by agents increase as the ϵ parameter increases, the negotiation process size also increases according to the ϵ parameter.
- **(b) Cooperation stage:** Agents in this stage become to hard in competing each other to acquire the large offer. This is because agents are hard to accept the large mistaken offer, even though they want to accept it, due to an increase of the randomness in stage (b) in comparison with stage (a). Since such a randomness also increases a probability of exceeding the last negotiation steps, the agents in this stage change to cooperate with the opponent agent from com pete each other, in order not to avoid exceeding the last steps. This contributes to decreasing the negotiation process size.
- **(c) Learning impossible stage:** Agents in this stage can no longer learn to avoid exceeding the last negotiation steps due to high randomness. Since most negotiations exceed the last negotiation steps, the negotiation process size increases again.

From this analysis, we can finally conclude that there is an applicable range for learning agent. In this simulation, the stage (a) is appropriate for the aim of the bargaining game.

5.4   Toward Guidelines for Modeling Agents

From the discussions in Sections 5.1 to 5.3, we can derive the following very rough guidelines for modeling agents:

1. **Cross-element validation:** Since even a slight difference (i.e., whether the m ax operation is used or not in an equation of updating Q-values) has an

essential influence on simulation results, it is indispensable to conduct the cross-element validation [19], instead of the cross-model validation such as "alignment of computational models" or "docking" [2]. The former validation is based on within-models, which compares simulation results of agents who differ only in one element (e.g., a learning mechanism in this research), while the latter validation is based on between-model, which compares simulation results of different computational models. Both approaches assert the importance of investigating whether different computational elements/models can produce the same results. However, key elements (e.g., the max operation), that determine characteristics of learning agents (e.g., risk taking or risk avoidance), are hard to be found by the cross-model validation but can be found by the cross-element validation. This is simply because the cross-model validation does not focus on an influence of detailed elements in computational models but analyze an influence of a structure of computational models. From this feature, the cross-element validation should be done to specify sensitive parts in multiagent-based simulation.

2. **Comparison of results between static and dynamic environments:** As described in Section 5.2, the negotiation process size in the static environment decreases, while that in the dynamic environment increases. Such a difference depends on whether an agent competes with an environment (i.e., the predetermined agent) or agents complete with each other. This clearly indicates that implications derived from simulations in the static environment (i.e., simulations using the predetermined agent) may not be useful because such simulation results differ from those in the dynamic environment (i.e., simulations using multiple agents) as is usual in social simulations. In this sense, we should conduct simulations in the dynamic environment, which is a very proper guideline. But, simulations in the static environment is useful to specify sensitive factors (i.e., the negotiation process size) by comparing with results in the dynamic environment. Since it is important to analyze such sensitive factors in detail in order for accurate implications, a comparison of results between the static and dynamic environments should be also done to determine candidates to be investigated in detail.

3. **Sensitivity analysis for specifying a range that agents perform well:** Since the dynamic environment has three stages in the negotiation process size (i.e., (a) competition; (b) cooperation; and (c) learning impossible), it is quite important to find such stage transition. This is because we have to clarify the stage that enables agents to acquire appropriate behaviors in order to conduct appropriate simulations. For this purpose, sensitive analysis is indispensable to clarify such applicable stage. In the bargaining game, the ϵ parameter should be set from 0 to 0.2 for an appropriate learning. What should be noticed here is that this type of analysis aims at specifying a range that agents perform well, which differs from the conventional sensitivity analysis that aims at finding sensitive parts in simulation results. From this different aim, we should also conduct sensitive analysis for specifying applicable range for learning agents.

6   Conclusions

This paper focused on how simulation results are sensitive to agent modeling in multiagent-based simulation and investigated such sensitivity by comparing results where agents have different learning mechanisms. Specifically, we employed two types of reinforcement learning, Q-learning and Sarsa, which differ very little (i.e., just one difference in updating Q-values). Through an analysis of simulation results in the bargaining game as one of the canonical examples in game theory, the following implications have been revealed: (1) even a slight difference (i.e., whether the max operation is used or not in an equation of updating Q-values) has an essential influence on simulation results. This indicates that cross-element validation is indispensable to specify key factors that affect simulation results; (2) testing in static and dynamic environments highlights the different tendency of results. Concretely, the different tendency is found in the negotiation process size not derive in the payoff, which suggests that the negotiation process size is sensitive factors in the bargaining game. Since it is important to analyze such sensitive factors in detail in order for accurate implications, a comparison of results between the static and dynamic environments is useful to determine candidates to be investigated in detail; and (3) three stages in both Q-learning and Sarsa agents (i.e., (a) competition; (b) cooperation; and (c) learning impossible) are found in the dynamic environment, while no stage is found in the static environment. Such finding suggests to conduct sensitive analysis for specifying applicable range for learning agents. In this example, the ϵ parameter should be set from 0 to 0.2 for an appropriate learning.

From these three implications, the following very rough guidelines for modeling agents can be derived: (1) cross-element validation for specifying key factors (e.g., the max operation) that affect simulation results; (2) a comparison of results between the static and dynamic environments for determining candidates (e.g., the negotiation process size) to be investigated in detail; and (3) sensitive analysis for specifying applicable range for learning agents (e.g., the parameter setting of ϵ). However, the above implications and rough guidelines have only been obtained from two learning mechanisms (i.e., Q-learning and Sarsa) and from one social problem (i.e., the bargaining game). Therefore, further careful qualifications and justifications, such as analyses of results using other learning mechanisms or in other domains, are needed to generalize our results. Such important directions must be pursued in the near future in addition to the following future research: (1) simulation mixing different kinds of learning mechanisms (preliminary results were shown in [20]); (2) simulation mixing agents who have different ϵ parameters; (3) simulation with more than two agents; and (4) investigation of the influence of the discount factor [14] in the bargaining game.

Acknowledgements

The authors wish to thank anonymous reviewers for useful, significant, and constructive comments and suggestions. The research reported here was supported in part by a Grant-in-Aid for Scientific Research (Young Scientists (B), 15700122) of Ministry of Education, Culture, Sports, Science and Technology (MEXT).

References

1. Axelrod, R. M.: *The Complexity of Cooperation: Agent-Based Models of Competition and Collaboration*, Princeton University Press, 1997.
2. Axtell, R., Axelrod, R., Epstein J., and Cohen, M. D.: "Aligning Simulation Models: A Case Study and Results," *Computational and Mathematical Organization Theory (CMOT)*, Vol. 1, No. 1, pp. 123–141, 1996.
3. Bäck, T., Rudolph, G., and Schwefel, H.: "Evolutionary Programming and Evolution Strategies: Similarities and Differences," *The 2nd Annual Evolutionary Programming Conference*, pp. 11–22, 1992.
4. Goldberg, D. E.: *Genetic Algorithms in Search, Optimization, and Machine Learning*, Addison-Wesley, 1989.
5. Güth, W., Schmittberger, R., and Schwarze, B.: "An Experimental Analysis of Ultimatum Bargaining," *Journal of Economic Behavior and Organization*, Vol. 3, pp. 367–388, 1982.
6. Holland, J. H., Holyoak, K. J., Nisbett, R. E., and Thagard, P. R.: *Induction*, The MIT Press, 1986.
7. Moss, S. and Davidsson, P.: *Multi-Agent-Based Simulation*, Lecture Notes in Artificial Intelligence, Vol. 1979, Springer-Verlag, 2001.
8. Muthoo, A.: *Bargaining Theory with Applications*, Cambridge University Press, 1999.
9. Muthoo, A.: "A Non-Technical Introduction to Bargaining Theory," *World Economics*, pp. 145–166, 2000.
10. Neelin, J., Sonnenschein, H., and Spiegel, M.: "A Further Test of Noncooperative Bargaining Theory: Comment," *American Economic Review*, Vol. 78, No. 4, pp. 824–836, 1988.
11. Nydegger, R. V. and Owen, G.: "Two-Person Bargaining: An Experimental Test of the Nash Axioms," *International Journal of Game Theory*, Vol. 3, No. 4, pp. 239–249, 1974.
12. Osborne, M. J. and Rubinstein, A.: *A Course in Game Theory*, MIT Press, 1994.
13. Roth, A. E., Prasnikar, V., Okuno-Fujiwara, M., and Zamir, S.: "Bargaining and Market Behavior in Jerusalem, Ljubljana, Pittsburgh, and Tokyo: An Experimental Study," *American Economic Review*, Vol. 81, No. 5, pp. 1068–1094, 1991.
14. Rubinstein, A.: "Perfect Equilibrium in a Bargaining Model," *Econometrica*, Vol. 50, No. 1, pp. 97–109, 1982.
15. Rummery, G. A. and Niranjan, M.: "On-line Q-learning Using Connectionist Systems", *Technical Report CUED/F-INFENG/TR 166*, Engineering Department, Cambridge University, 1994.
16. Ståhl, I.: *Bargaining Theory*, Economics Research Institute at the Stockholm School of Economics, 1972.
17. Sutton, R: "Generalization in Reinforcement Learning: Successful Examples Using Sparse Coarse Coding," In Touretzky, D. S., Mozer, M. C., and Hasselmo M. E. (Eds.), *Advances in Neural Information Processing Systems*, The MIT Press, pp. 1038–1044, 1996.
18. Sutton, R. S. and Bart, A. G.: *Reinforcement Learning – An Introduction –*, The MIT Press, 1998.

19. Takadama, K., Suematsu, Y. L., Sugimoto, N., Nawa, N. E., and Shimohara, K.: "Cross-Element Validation in Multiagent-based Simulation: Switching Learning Mechanisms in Agents," *The Journal of Artificial Societies and Social Simulation (JASSS)*, Vol. 6, No. 4, http://jasss.soc.surrey.ac.uk/6/4/6.html, 2003.
20. Takadama, K. and Fujita, H.: "Lessons Learned from Comparison Between Q-learning and Sarsa Agents in Bargaining Game," *NAACSOS (North American Association for Computational Social and Organizational Science) Conference 2004*, 2004.
21. Watkins, C. J. C. H. and Dayan, P.: "Technical Note: Q-Learning," *Machine Learning*, Vol. 8, pp. 55–68, 1992.

Agent-Based Modelling of Forces in Crowds

Colin M. Henein[1] and Tony White[2]

[1] Institute of Cognitive Science, Carleton University,
1125 Colonel By Drive, Ottawa, Ontario, K1S 5B6, Canada
cmh@ccs.carleton.ca
[2] School of Computer Science, Carleton University,
1125 Colonel By Drive, Ottawa, Ontario, K1S 5B6, Canada
arpwhite@scs.carleton.ca

Abstract. Recent events have highlighted the importance of good models of crowds, however many existing crowd models are either computationally inefficient, or are missing a crucial human behaviour in crowds: local pushing. After discussing some essential aspects of force in crowds, and considering some existing models, we propose an efficient agent-based model of crowd evacuation that incorporates pushing forces and injuries. Basing our model on existing work, we extend this model to investigate force effects at different crowd densities. Analysis of our model shows significant effects of force on the crowd, as well as significant effects of crowd density when measuring the number of agents still trapped inside the space after a fixed time.

1 Introduction

Crowds are a part of our everyday lives. While most crowds are safe, some can be dangerous: this year alone over 250 pilgrims were crushed during the Hajj, and over 80 were crushed at the lantern festival in Beijing. Crowd effects are not limited to these exterior settings; for example, architects designing large venues (e.g. arenas and lecture theatres) need to understand crowd exit behaviours. Thus, good models of crowds can be very helpful to many professionals.

From a modelling perspective, crowd events are interesting to social simulation researchers because their associated phenomena are largely emergent in nature. Interesting crowd behaviours with which we are familiar include the formation of rivers of movement through otherwise stationary crowds, spontaneous formation of lanes when pedestrians are moving in opposing directions, and roundabouts when paths cross [1]. Crowds also demonstrate speed and force-related effects, including rainbow-like arching structures as pedestrians jam and clog at exits, bursty exit rates as jostling prevents smooth use of doors, and inability for crowds to pass through each other when speeds (and forces) are high [1]. All these crowd effects are generally observable from an overhead perspective, and we are all aware from our personal experience that individual pedestrians do not plan them.

Force effects are particularly important in modelling crowd behaviours. Although people generally try to move toward goals, force effects can cause them to be pushed away from their desired trajectories and accurate models must reflect this. Also, the

P. Davidsson et al. (Eds.): MABS 2004, LNAI 3415, pp. 173–184, 2005.

presence of crowd members injured by excessive force can significantly affect the ability of others to move freely. In an evacuation situation, increased desired walking speed leads to increased forces, and these forces tend to cause additional delays to those trying to exit. Models that do not represent pushing forces therefore cannot directly account for all these additional delays.

The purpose of the present research was to create an efficient, agent-based, individual-centred model of crowd evacuation that incorporates pushing forces and injuries. In the remainder of this section we will consider existing models of crowd behaviour (including some agent-based ones), and discuss their advantages and disadvantages; ultimately we will conclude that one promising model could be improved by the addition of force. The following sections will outline this model, essential aspects of force in crowds, and the details of our implementation of these phenomena. Finally, we will provide an analysis of our results as well as future directions for this research.

1.1 Existing Models of Crowd Behaviour

Many explanations of crowd behaviour have been proposed over the last forty years [review: ref. 1]. Historically, many of these investigations have borrowed physical analysis techniques (e.g. fluid dynamics) that are focussed on aggregate factors. This approach obscures individual-level interactions (e.g. collision avoidance and pushing) which may be important for accurate modelling, and which are of particular interest to researchers seeking to study individual movement and safety.

Recently, agent-based models have been proposed which begin to address this issue by operating at the level of individuals. A frequently cited model of this type is the social forces model, advanced by Helbing and colleagues [2]. A strongly physicalist model, it calculates forces acting on agents to determine movement, with excessive forces leading to agent injuries. The model considers the effect that each agent has upon all the other agents, almost as if the model were a simulation of an n-body problem in astrophysics. Physical forces (e.g. friction encountered when brushing past another person, or elastic force due to body compressions) are modelled, as are social forces (desire to change direction to avoid another). One problem with Helbing's model is that of computational complexity. Simulation update is $O(n^2)$ due to the calculation of the effect that each agent (and obstacle) has on all the other agents. This may limit the model's ability to simulate many agents.

A more efficient model may be obtained by avoiding the direct consideration of every agent's effect on all the others. We call these models local-interaction models. Cellular automata models take this approach, as do agent-based models that rely on local information to simulate an unfolding crowd scenario.

Kirchner and colleagues have proposed a model along these lines that emphasizes decentralised processing by independent agents [3]. Agents representing crowd members are distributed on a grid that serves not only as a coordinate system to track agent movement, but also as a data structure storing location-specific data of use to agents. As each agent calculates its next step, references to physical features of interest in the environment (e.g. doors) are replaced by references to a value stored on the grid cell in question (e.g. distance from here to nearest exit).

One deficiency of many local-interaction models is that, unlike Helbing's model, they do not address the question of force applied by agents to other agents. Individuals in Kirchner's crowds, for example, are well behaved as they await their opportunity to exit. This means that the model cannot account for delays in exiting due to pushing or the secondary obstruction effects of injuries. We believe that results can be improved by creating an agent-based local-interaction model incorporating force.

2 Starting with an Existing Model: Kirchner

Instead of creating an entirely new local-interaction model demonstrating force effects of interest we have elected to base our model on the Kirchner model [3]. This is because Kirchner's model is already capable of demonstrating many of the effects discussed in the introduction; furthermore it provides a computationally efficient and extensible base on which to construct force effects.

In this section we will describe the basic model. In the next section, we describe how pushing and inter-agent forces can be integrated into the model.

2.1 Floor Field Modelling

The modelled space is divided into square grid cells. It makes use of *floor fields* to provide location-specific data to individual agents in the model. Each floor field defines a set of values, one value for each cell. An individual agent has access to the floor field values on its current cell, as well as those on the four cardinal neighbouring cells.

The Kirchner model defines two floor fields: the *static* field and the *dynamic* field. The static field is a gradient that indicates the shortest distance to an exit. Thus, an agent is in a position to know the distance to the closest exit by consulting the static field value on its current cell; by consulting the values in neighbouring cells the agent can determine the *direction* in which to move to reach an exit.

The dynamic field is a measure of agent movement. An agent increments the dynamic field level when moving. By analogy with ant pheromones [4], the dynamic field diffuses and evaporates. By consulting the dynamic field an agent can follow other nearby agents without directly considering the position of any other agent.

By varying the degree to which an agent is sensitive to one field or the other, this model can simulate different pedestrian strategies within the crowd, as well as situations in which pedestrians have less than perfect knowledge of the location of exits. For example, to model herding behaviours in anxious crowds, the agents' sensitivity to the dynamic field can be increased, thereby promoting agent interaction. To model reduced visibility in an unfamiliar environment agents' sensitivity to both fields can be decreased.

2.2 Update Rules

Kirchner's model proceeds by iterating a set of basic steps, represented graphically in figure 1. First (step K1), a score – representing desirability – is calculated for each cell according to the following formula:

$$Score(i) = \exp(k_d D_i) \times \exp(k_s S_i) \times \xi_i \times \eta_i \tag{1}$$

where $Score(i)$ represents the score at cell i; D_i is the value of the dynamic field in that cell and S_i is the value of the static field in that cell; k_d and k_s are scaling parameters to the model governing the degree to which an individual is aware of other-agent movement and location of exits, respectively; ξ_i is 0 for forbidden cells (e.g. walls, obstacles) and 1 otherwise; η_i is 0 if an agent is on the cell, and 1 otherwise.

Having determined the scores, each agent then uses them (step K2) to probabilistically determine the cell that it would like to move onto in the next time step (motion is permitted to the four cardinal neighbours, and 'motion' to the current cell is also possible). The agent calculates the probability of moving from cell i to a neighbouring cell j, by dividing the score for j by the sum of the scores of those cells adjacent to i:

$$p_{ij} = Score(j) \left(\sum_{a \in N(i)} Score(a) \right)^{-1} \tag{2}$$

where p_{ij} is the desired probability and $N(i)$ gives the set of cells that are neighbouring cells to i, including i itself.

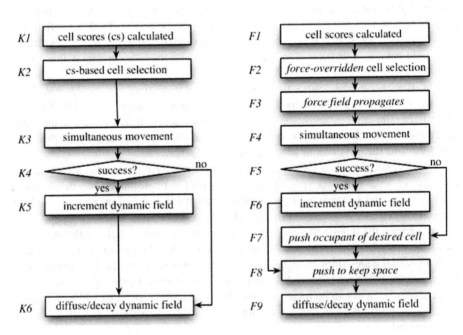

Fig. 1. (a) Update rules in Kirchner (*left*) and (b) force-modified model (*right*). Italicised rules highlight differences with the Kirchner model

When all agents have decided on their desired grid cells, they all move simultaneously (step K3) and increment the D_i value of their original cell by one (step K5). Agents selecting their own cell do not move, and do not update the dynamic field. As no two agents may occupy the same space, any conflicts for moving agents (step K4) are resolved by allowing a random agent to occupy the desired cell, if empty; the other conflicting agents do not move.

Finally, D_i is diffused and decayed (step K6). To diffuse, a cell distributes a proportion of its D_i value equally among its 4 cardinal neighbour cells. To decay, the cell discards a proportion of its D_i value. The decay and diffusion proportions are input parameters to the model.

Agents moving onto a cell designated as an exit cell are removed from the model upon the completion of the time step, and are deemed to have exited at that time.

3 Modelling the Application of Force

Force is an important aspect of crowd models. Forces in crowds affect the ability of pedestrians to exit in a timely manner. A model without force may not accurately model pedestrian exit behaviours. In this section we add force to Kirchner's model.

3.1 Aspects of Force

Before proceeding to create a model that includes forces it is worth mentioning five essential properties of crowd forces that will guide our design.

First, people in crowds do not push randomly; pushing occurs for a reason. People push in a particular direction when they want to move in that direction and are prevented from doing so. Also, people push when they want to maintain their personal space in a crowd.

Second, forces *propagate through* crowds; some researchers have reported that force can move through a crowd like a shockwave [5]. It is essential to note this fact when modelling force, as it dictates that force applied by individuals in one part of a modelled crowd must propagate through that crowd with certain time characteristics; pushing and force propagation in crowds is not instantaneous.

Third, forces are directed and any model should consider force to be a vector rather than a scalar.

Fourth, from a modeller's perspective there are two important effects of force on individuals within crowds: Individuals subjected to extremes of force become injured; these injuries have consequences for the injured agent (immobility, inability to exit) but also for others in the vicinity (injured people are obstacles to others). The other important effect of force is that when individuals are subjected to non-injuring (smaller) forces, these affect movement. Therefore a modelled pedestrian who is pushed ought to move in the direction of the force, rather than in some other direction that they might prefer.

Fifth, is location-specificity. When many people in a crowd are pushing together in a similar direction then the applied force combines additively to have a large effect on those receiving the force. Often, injuries tend to occur near the front of crowds, as the aggregate force from behind becomes overwhelming. Some reports have concluded that injuries have occurred towards the centre of a crowd, where the force from those pushing off the wall at the front met with force generated by those pushing from the rear [5]. In any case, modelled injuries should not be randomly distributed throughout the modelled space, but should instead be produced at physically reasonable positions within the crowd.

3.2 Adding Pushing to Kirchner's Model

When adding force to the model, we have been conscious of the performance advantages of the floor field design. Accordingly, force is represented in the model as a third floor field.

The force field value on each cell represents the force experienced by the agent on that cell in the current time step. In order to represent the directionality of force, we have used a vector field instead of the scalar fields used by Kirchner. We define a vector floor field to hold both a magnitude and direction for each cell in the model, rather than the scalar floor fields that hold only a scalar value for each cell. Where applicable, we use vector addition to update the vector field.

Agent Rule Modifications. The agent applies force in two circumstances (see Figure 1b). If the agent wishes to move into a cell, but is blocked (another agent has moved there first, or an occupying agent is present) then the agent will push the occupant of the desired cell; in other words, the agent will vector-add its pushing force (parameter k_{push}) into the force field value on the desired cell, in the direction of desired travel (figure 1b, step F7).

In the second circumstance of force application, the agent applies force to neighbouring agents in order to maintain its space in the crowd (step F8); one quarter of the pushing force ($k_{resist} = 0.25 \ k_{push}$) is vector-added into each adjacent occupied cell in the direction of that cell.

The agents follow Kirchner's movement rules, with the following exception: If the force experienced by an agent (i.e. the value of the force floor field at the agent's location) exceeds the force ($f_{divert} = k_{push} + k_{resist}$) that it can apply to another agent, then the normal score-based probabilistic cell selection described above is bypassed (step F2). Instead, the agent is forced to select the cell in the direction of the force. Thereafter, normal movement rules apply (only one agent can actually move onto that cell, blocked agents apply force, etc).

Force Update Rules. The force field is an active field; like the dynamic floor field that tracks agent movement, the force field propagates through the model according to certain rules (step F3). The basic force rule is that as a part of each time step, the entire force on a particular cell moves on to the next cell (in the direction of the force vector). Thus, force is propagated through the model over time. Forces that collide on a cell are vector added together. Force cannot be propagated through empty space; if force becomes stranded on an empty space its magnitude is reduced to zero. Likewise, walls and obstacles absorb force, and do not re-transmit it.

Injury Behaviours. Agents become injured if the sum of forces acting on them reaches a threshold level ($f_{injuring}$), which is an input parameter to the model. The scalar sum of incoming forces is used, rather than the vector sum for this purpose. Once injured, an agent ceases movement; its cell is designated as an obstacle cell (like walls).

3.3 Controlling Crowd Density

Force transmission within a crowd is highly dependent on crowd density because force cannot travel through empty space. We found that Kirchner's model created crowds that were highly dense (virtually all agents packed around the door, few empty grid squares). Some crowds in the world may be like this, but most are not; generally there are tightly-packed areas within crowds (e.g. near the exit) and more space within the crowd at other points (e.g. near the back). We wanted to evaluate our force model in the context of differing crowd densities.

Accordingly, we further modified Kirchner's model to allow for a density control. This control was achieved by refining the η term of the cell scoring equation (1). It was also necessary to exclude the agent's current cell from the neighbourhood for the probabilistic cell selection of equation (2).

Regarding the η term: Since movement within the model is simultaneous, one agent should very well be able to leave a cell c at the same time as another agent is entering it. The η term prevents this, however, by stopping agents from selecting cell c if it is occupied (by dropping its probability of selection to 0), meaning that no agent will try to move into c as its current occupant leaves; this guarantees that any occupied cell will be unoccupied for at least one time step when its occupant leaves.

To resolve this problem we have redefined the occupied value of the η term as an input parameter to the model, k_n. When a cell is unoccupied then its η is set to 1, and when a cell is occupied then its η set to the value of k_n. This means that agents can now (with a certain probability) select occupied cells – in effect betting that the occupying agent will leave. (Note that the rules regarding multiple agents on the same cell are still in force, so an agent does not move if it selects a cell that does not then become unoccupied.)

Regarding the agent's neighbourhood: If the agent's current cell is not excluded from equation (2), then the η term manipulation does not yield a good density control as the probability of remaining on the current cell overwhelms the scaled probability of moving to a neighbouring cell.

When k_n is high then a dense crowd should be obtained because agents select cells irrespective of their occupancy; in a packed crowd we expect most agents will stand still waiting for the neighbouring cell with the most attractive score. When k_n is low then a lower density crowd should emerge because agents are encouraged to move to unoccupied cells, and this movement within the crowd should preserve space. Thus, this new parameter should allow us to model situations ranging from closely packed crowds ($k_n = 1$) to sparse crowds ($k_n = 0$).

4   Results

The model was implemented in NetLogo 2.0 [6] (on Mac OS X 10.3.2, Java 1.4.2).

The simulated floor space was 31×31 cells in size, not including the wall cells surrounding the floor space. The results presented here are for one exit cell situated in the middle of one wall, with no other obstacles (figure 2).

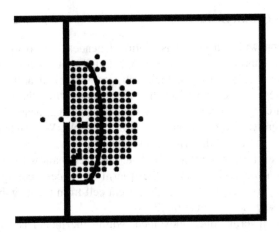

Fig. 2. Snapshot of representative model execution. Agents (*circles*) clustered around door (*white cell in black walls*) in typical arch formation. Injured agents (*black squares*) are present due to previous pushing. Individuals within the half-crowd closest to door (*overlaid curve* added for illustration) are most likely to become injured

Two hundred individuals were assigned random starting cells within the space. Initially the force field had a magnitude of 0 on every cell. The static floor field was initialized according to the linear distance between the centre of each cell and the centre of the exit cell. For the purposes of measuring the effects of force on the model, k_s was set to 10 and k_d was set to 0. The agent's pushing force, k_{push}, was set to 1, and the agent's injury threshold, $f_{injuring}$, was set to 23.

We report performance of the model in terms of the number of agents that remain in the space after a fixed number of iterations. (Although it may seem more natural to report the number of iterations required for all agents to exit, this measure is problematic because if an injury occurs such that an agent blocks the exit it prevents the remaining agents from ever exiting.)

4.1 Effect of Varying k_n

To investigate the effect of introducing the k_n parameter we measured our model running with Kirchner's rules as a base case. We then removed agents' own cells from their neighbourhoods in equation (2), and measured the result of setting the k_n parameter to values of 0, 0.5 and 1.0.

We observed that upon introducing the 4 cell neighbourhood more agents were able to exit. For the 4 cell neighbourhood, a sparse crowd was observed when $k_n = 0$, while arching structures with only a few empty cells are obtained when $k_n = 1$. Exits when $k_n = 0$ occur at quite a regular rate, not faster than 1 exit per 2 timesteps. When k_n is increased to 0.5 this rate increased, remaining regular.

Numerical result are provided in Table 1.

Table 1. Agent escape statistics with Kirchner 5-cell neighbourhood vs. 4-cell neighbourhood with varying k_n (means and standard deviations of 10 runs). All differences are significant with a two-tailed Student t-test ($p < 0.0001$) except between the pair*

neighbourhood	k_n	# remaining after 350
5-cell	-	75.3 (6.3)
4-cell	0.0	55.1 (3.7) *
4-cell	0.5	28.7 (5.7)
4-cell	1.0	57.7 (4.3) *

4.2 Effect of Adding Force Rules

We measured the effect on agent escape when force was added to the model at the same values of k_n that were tested above.

Numerical result are provided in Table 2.

We observed that forces did propagate through the crowd, gaining strength as multiple agents joined in the pushing action. Forces tend to propagate both into the crowd (towards the vertical centre and towards the door) and out of the crowd (from the vertical centre out) as agents push in the direction they want to move, and simultaneously resist those around them.

Table 2. Agent escapes and injuries with revised model (means and standard deviations of 10 runs). All differences significant with a two-tailed Student t-test ($p < 0.05$)

k_n	# remaining after 350	# injured after 350
0	66.4 (4.7)	0.0 (0.0)
0.5	80.9 (11.1)	4.7 (2.1)
1.0	105.4 (30.8)	7.1 (2.0)

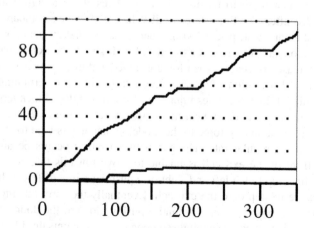

Fig. 3. Number exited (*upper line*) and injured (*lower line*) with time during a representative model execution

Injuries in the model tend not to occur right at the beginning or end of the simulation, but rather begin when a significant crowd density has developed near an exit, and end when enough agents have exited to relieve some pressure. In our trials, injuries tended to occur between time steps 50 and 150 (e.g. figure 3). Some injuries occurred in groups of 2-4 nearby cells, while other injuries were isolated. The majority of injuries occurred in the half-crowd closest to the door (see figure 2), and tended to be distributed roughly along an arch that passed near the centre of the crowd (at the time the injury occurred). Agents near the wall with the door, but towards the edges of the arch structure were also frequently affected by injuries. We also observed a more bursty exit rate than in trials without force.

5   Discussion

The modification of Kirchner's neighbourhood to a 4 cell neighbourhood decreased the number of agents remaining in the space after 350 iterations of the model (table 1). This is due to an increased movement rate because agents could no longer choose to spend time standing still.

Revision of Kirchner's cell-selection (scoring) function by introducing a value of $k_n = 0.5$ significantly decreased the number of agents remaining after a fixed time (table 1) compared to other values of k_n. This is because the revised cell-selection rule no longer requires that occupied cells spend a time step unoccupied. Since this rule applies to the exit cell as well, the original model's maximum exit rate was one agent per two time steps, while the new rule allows for a maximum exit rate of one agent per time step. In addition, the new rule increases crowd density, placing more agents in the vicinity of the door, and thereby in a position to readily exit when possible.

At $k_n = 1.0$ agents completely ignore the occupancy of a cell when rating its desirability. This may be a poor strategy in dense crowds as the occupying agent may not be able to move. When k_n is 0.5 agents will select occupied cells but still prefer unoccupied ones – a better trade-off according to table 1.

The addition of force to the model (table 2) results in a significant increase in agents stranded after 350 time steps at all values of k_n when compared with table 1 (two-tailed Student t test, $p < 0.0001$). Hence, the simulation bears out the fact that pushing (paradoxically) increases exit times [1]. It should be noted that the standard deviations increase noticeably when force is added and injuries occur. This is because the number and distribution of injuries play a large role in determining the effect of force on the model. In some cases injuries in front of the door block large numbers of agents, while injuries that occur near the periphery have minimal effect.

Another effect of adding force to the model is that agents tend to exit in bursts instead of at a fairly regular rate. The gap between bursts occurs because an agent a standing in front of the exit cell is unable to move towards that cell; this is due to other agents lined up to the left and right of a pushing a in a different direction, overriding a's desire to step onto the exit cell. Eventually the forces acting on a may be disrupted, or overcome by agents behind a pushing toward the door, and a will exit. This creates a hole that can be capitalised upon by other agents until forces build up to once again jam the door. This bursty exit pattern has been reported as a hallmark of the arches that form at exit points [1].

Bursty exit characteristics may be precluded if injuries occur near, but asymmetric to the door. In these cases, the bursting behaviours are extinguished because the cell with the injury acts as a force break, taking up the force and helping to isolate agents nearer to the door from the paralysing force scenario just described. Force relief by obstacles is a known phenomenon in crowd research [1]. Its effectiveness in regulating bursty output in this model is an effect that was not programmed into the rules; rather it emerges from the effects of force on agents and obstacles. Its emergence in the model is further evidence that our force paradigm is a promising one.

Because crowd density increases with increasing k_n, and crowd density should lead to more pushing and therefore delay and injury, we can further test our model by predicting that delay and injury should increase with k_n when forces are applied by agents. In fact, this prediction is borne out (table 2). Crowd density directly affects the number of injuries because fewer empty cells within the crowd mean more effective force propagation. The increase in density is accompanied by an increase in the number of agents left inside after 350 time steps, both because the number of injuries increases, and because greater force build up results in longer intervals between bursts of exiting agents. Note that when k_n is 0 no injuries were obtained because spacing in the crowd prevented the transmission and build-up of force.

The principle of location-specificity described above is observed in the model. Generally consistent injury positions re-occur across multiple simulation tests. It is a consequence of the generally semi-circular arching structure that there are more agents pushing on the chord parallel and adjacent to the wall, than on the chord parallel to the wall but near the back of the crowd. The combined pushing power of the larger group of agents along the wall, mostly directed parallel to the wall towards the door, means that injuries tend to occur near this wall. Forces tend to concentrate near the door (result of agents pushing towards the door) and towards the edges of the crowd (result of agents pushing back to maintain space in the crowd).

6 Conclusion

We have described an agent-based model of crowd dynamics based upon local interactions with floor fields. After the addition of a force floor field, the model continued to demonstrate the characteristic arching structure that we expect to see in crowd evacuation simulations. Moreover, our new model is capable of answering quantitative and qualitative predictions related to crowd injuries and egress dynamics.

A flexible density control was successfully added to Kirchner's model by turning his occupancy-exclusion parameter η into a more flexible density control: k_n , and by modifying his neighbourhood structure.

Considerable work remains in the development of this model. In particular, the model needs to be related to real world data, and calibrated so that units of force and distance in the model can be related to real units of force and distance, and so force capability and thresholds are related to human capacities. Finally, a quantitative analysis of the bursting exit behaviours, and role of the k_n parameter in regulating crowd density, needs to be undertaken.

References

1. Helbing, D., Farkas, I.J., Molnár, P., Vicsek, T.: Simulation of pedestrian crowds in normal and evacuation situations. In: Schreckenberg, M., Sharma, S.D. (eds.): Pedestrian and evacuation dynamics. Springer-Verlag, New York (2002) 21-58
2. Helbing, D., Farkas, I., Vicsek, T.: Simulating dynamical features of escape panic. Nature v. 407, 28 September 2000. 487-490
3. Kirchner, A., Schadschneider, A.: Simulation of evacuation processes using a bionics-inspired cellular automaton model for pedestrian dynamics. Physica A 312 (2002) 260-276
4. Bonabeau, E., Dorigo, M., Theraulaz, G.: Swarm intelligence: From natural to artificial systems. Oxford university press, New York (1999)
5. Fruin, J.: The causes and prevention of crowd disasters. In: Smith, R.A., Dickie, J.F. (eds): Engineering for crowd safety. Elsevier, New York (1993)
6. Wilensky, U.: NetLogo. Center for connected learning and computer-based modeling, Northwestern University, Evanston IL. 1999. http://ccl.northwestern.edu/netlogo

An Investigation into the Use of Group Dynamics for Solving Social Dilemmas

Tomohisa Yamashita, Kiyoshi Izumi, and Koichi Kurumatani

Cyber Assist Research Center (CARC),
National Institute of Advanced Industrial Science and Technology (AIST),
Aomi 2-41-6, Koto-ku,
Tokyo 135-0064, Japan
{tomohisa.yamashita, k.kurumatani}@aist.go.jp
kiyoshi@ni.aist.go.jp

Abstract. In this research, we propose some group dynamics that promote co-operative behavior in systems with social dilemmas and hence enhances their performance. If cooperative behavior among self-interest individuals is established, effective distribution of resources and useful allocation of tasks based on coalition formation can be realized. In order to realize these group dynamics, we extend the partner choice mechanisms for 2-IPD to that for N-person Dilemma game. Furthermore, we propose group split based on metanorm as a new group dynamic. A series of evolutionary simulations confirm that this group dynamic: i) establishes and maintains cooperation, and ii) enhances the performance of the systems consisting of self-interest players in Social Dilemmas situations.

1 Introduction

Recently, multiagent systems have applied more and more frequently as a framework for constructing large distributed systems. The introduction of autonomous collaborating agents gives more flexibility and efficiency to systems. In multiagent systems, usually the agents have only incomplete information and limited ability to solve problems in the environment [9, 19]. Faced with a problem it cannnot solve, an agent may seek to get together with other agents and form a collaborative group. The question is then whether cooperative behaviours will result from such a group. It is difficult to promote cooperative behaviors in the case that an individual agent can acquire a higher reward by non-cooperative behavior even though cooperative behaviour by all would maximise the reward of all [5, 6, 8, 10, 15, 16]. This situation is generally called a Social Dilemma [11], where free-riders who choose non-cooperative behaviors decrease the total reword in the group. In multiagent systems, this problematic situation is often observed in resource distribution and task allocation without central authority. Self-interest agents do not choose a e cooperative behavior for maximizing the reward of the group because they prefer the individually best outcome to the collectively best outcome. If some decentralised mechanisms ca be devised to prevent free-riders from joining cooperative groups, the total reward in the group can be increased and the performance of the systems can be enhanced. Therefore group dynamics which do not require the existence of

P. Davidsson et al. (Eds.): MABS 2004, LNAI 3415, pp. 185–194, 2005.

central authority, are one of the most important mechanisms promoting cooperation in multiagent systems.

In this paper, in order to realize effective group dynamics in dilemma situation, we use the mutual choice mechanisms [1, 18, 20] as the basic interaction for group formation, which is the matching mechanism in a multiple 2-person Prisoner's Dilemma (2-PD). Some previous researchs revealed that the conventional partner choice mechanisms for matching two persons can promote the establishment of coopeartion [7, 12, 14, 17]. Here a more general dilemma situation is investigateed, the iterated N-person Prisoner Dilemma game. Furthermore, we propose a group split rule based on the concept of metanorm [2]. The effect of our proposed group dynamics on the enhancement of system performance is confirmed through an agent-based simulation with evolutionary approach.

2   Group Dynamics

In this paper, we use mutual choice for the matching mechanism in multiple 2-P PD games, to show how effective group dynamics can be in dillema situations.[1] Group dynamics in the dilemma situation are modeled as a 2 stage game. The first stage is where agents choose group members. The groups are determined through some group dynamic mechanism, – a population of all agents is partitioned. A group is defined as a subset of the overall agent set. Each agent can join only one group at any one time. The order of decision-making by the agents is set as random. According to this order, the agents make decisions one by one, so that groups are gradually formed. The second stage is a dilemma game. In the groups consisting of two or more agents, agents play the N-PD with their group members. The result of the N-PD in a group is independent of the agents in other groups. Where groups consist of a single agent, that agent acquires a fixed payoff.

2.1   Unilateral Choice

In group formation based on unilateral choice [13], agent i can join group k surely, i.e., group k cannot refuse agent i. Each agent has the alternative of forming a new group or joining an existing group. Agent, i, chooses a group, k, which is the most tolerable to it. Agent i is then added to group k. If there is no tolerable group for agent i, agent i makes no offers and forms a new group.

2.2   Mutual Choice

In group formation based on mutual choice, the agent making an offer and the group receiving it form a new group only if both agree. Therefore, a group has the possbilities of either refusing or accepting an offer. After agent i makes an offer to group k (this process is the same as that for unilateral choice.), group k can decide to refuse or accept agent i by majority vote based on the decision of all members. If the majority of agents in group k agree to accept player i, group k accepts the offer of agent i and then agent

---

[1] For more details see Yamashita and Ohuchi[21].

i joins group k. If group k refuses agent i, agent i makes an offer to the second most tolerable group. Agent i continues making offers until a group accepts its offer or until all groups tolerable to it refuse its offer. If agent i is refused by all tolerable groups, agent i forms a new group.

2.3 Group Split

A group split rule is proposed based on the concept of a metanorm. According to Axelrod [2], if there is a certain norm, a metanorm is to "punish, not only against the violators of the norm, but also against anyone who refuses to punish the defectors." In this model, a norm based on mutual choice is "don't choose defectors as group members." A metanorm based on mutual choice is "don't choose, not only defectors, but also anyone who choose defectors as group members." This metanorm effectively realizes group split by dividing the group of agents agreeing to the acceptance of an agent (agreeing agents) and the other group opposing this (opposing agents). From the point of view of the opposing agents, the agreeing agents violate the metanorm because the agreeing agents choose the agent that opposing agents consider to be a defector. In order not to choose the agents who choose defectors as group members, the opposing agents leave from the group.

2.4 Re-offering

By the introduction of the rule of group split, the number of groups of only one agent may increase. Re-offering is proposed as a mechanism to increase the chance of creating a group of several agents as soon as possible.

2.5 Dilemma Game

In this model, we use a more general dilemma game than the N-PD in [6]. After the group dynamics, the players in each group play the dilemma game with group members if the players are in groups of more than two players. Otherwise, a player in a group only consisting of itself acquires the fixed payoff for lone players, the *reservation payoff* $P_{reservation}$, instead of the payoff of the dilemma game [1, 18, 20]. Each group member decides its contribution to its group, and then the profit given by the total contibuted by all members is redistributed equally to the group members. Each player i in group k contributes some amount, $x_i \in \{0, 1\}$, to group k as the strategy, and has a payoff function, F_i. The total contribution of all players in group k amounts to $X \equiv \sum_{i \in G_k} x_i$. The payoff function of player i in group k can be written as

$$F_i(x_i; X, |G_k|) = a\frac{X}{|G_k|} + b(1 - x_i) \tag{1}$$

where $|G_k|$ is the size of group k, and a and b are positive constants.

3 Simulation

In our simulations, a genetic algorithm (GA) is applied to evolve the player's strategies [3, 4]. The two dimensions of a strategy, cooperativeness C_i and vengefulness V_i, are each divided into $2^x - 1$ equal levels, from 0.0 to 1.0. Because $2^x - 1$ levels are represented by x binary bits, a player's strategy needs a total of $2x$ bits: x bits for cooperativeness C_i and x bits for vengefulness V_i.

Each generation consists of an iteration of group formation and split processes, and then the Dilemma game. At the beginning of the GA, each player's strategy in a population is assigned a fitness equal to its average payoff given per payoff received. Uniform crossover is applied to the strategies of a player and a partner to obtain a new strategy for one offspring if the fitness of the partner is better than that of the player in tournament selection.

Since our purpose is to examine whether the group dynamics we investigate can promote cooperative behavior among players and enhance the performance of systems, we pay attention to the development of players' strategies and the average payoff of all players. In order to confirm the effect of the proposed group dynamics, we compare four settings: case 1) only group formation based on unilateral choice, case 2) only group formation based on mutual choice, case 3) group formation based on mutual choice and group split, and case 4) group formation based on mutual choice, group split and re-offering. We define the establishment of cooperation as the situation where both the average cooperativeness of all players ($\overline{C}$) and the average vengefulness ($\overline{V}$) are bigger than 0.8. The important parameters are shown in Table. 1.

3.1   Establishment of Cooperation

The number of times cooperation is established in 40 trials of the four cases is 0 in case 1, 4 in case 2, 12 in case 3, and 40 in case 4. In all trials of case 1, there was little cooperativeness and vengefulness, i.e., cooperation was not established at all within 5,000 generations. In 36 trials of case 2, cooperation was not established within 5,000 generations. In the remaining 4 trials, there were great deal of cooperativeness and vengefulness, i.e., cooperation was established. In 28 trials of case 3, cooperation was not established within 5,000 generations. In the remaining 12 trials, cooperation was established. In all trials of case 4, cooperation was established.

Three typical developments of the average cooperativeness and vengefulness of players are shown in Figs. 1, 2, and 3. In these graphs, the horizontal axis represents the gen-

Table 1. Common parameters in the simulations

Number of players	50
Number of generations	5000
Number of group dynamics per generation	200
Coefficient of payoff function a	1.0
Coefficient of payoff function b	0.6
Payoff for lone player $P_{reservation}$	0.1
Initial value of expected cooperation π	1.0
Initial value of Cooperativeness C_i	0.0
Initial value of Vengefulness V_i	0.0
Mutation rate	0.05
Binary bits for C_i and V_i (total bits)	10

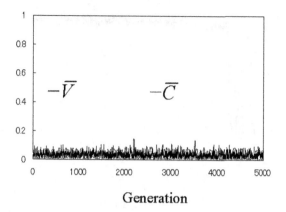

Fig. 1. Example of the failure of the establishment of cooperation in case 1: the average cooperativeness $\bar{C}$ and vengefulness $\bar{V}$ from 0 to 5,000 generations

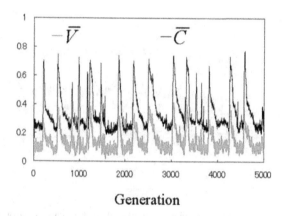

Fig. 2. Example of the failure of the establishment of cooperation in case 2 and 3: the average cooperativeness $\bar{C}$ and vengefulness $\bar{V}$ from 0 to 5,000 generations

eration, and the vertical axis represents the average cooperativeness and vengefulness of players. The typical behavior of $\bar{C}$ and $\bar{V}$ in case 1 for 5,000 generations is shown in Fig. 1. Throughout the generations, $\bar{C}$ and $\bar{V}$ fluctuated in the range of 0.0 to 0.1. The typical behavior of $\bar{C}$ and $\bar{V}$ in case 2 and 3 when cooperation was not established is shown in Fig. 2. Usually, $\bar{C}$ remained in the range of 0.0 to 0.2 and $\bar{V}$ remained in the range of 0.2 to 0.4. However, occasionally $\bar{C}$ rose to 0.4 and $\bar{V}$ fell to 0.3, and then $\bar{C}$ and $\bar{V}$ returned to their initial states. This fluctuation was repeated throughout the generations. The typical behavior of $\bar{C}$ and $\bar{V}$ in case 2, 3, 4 when cooperation was established is shown in Fig. 3. Once $\bar{C}$ and $\bar{V}$ reached 1.0, this state continued to remain there and did not transfer to another state.

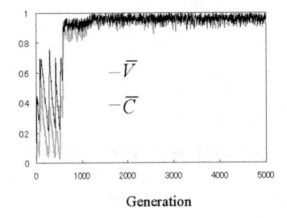

Fig. 3. Example of the establishment of cooperation in case 2, 3, 4: the average cooperativeness $\bar{C}$ and vengefulness $\bar{V}$ from 0 to 5,000 generations

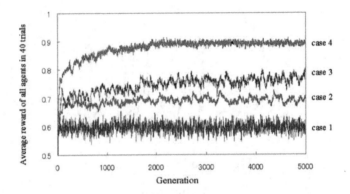

Fig. 4. The average payoff of all players in 40 trials of four cases of group dynamics

3.2   Average Payoff

The average payoffs of all players in 40 trials of four cases are ranked in descending order as case 4, 3, 2, 1. In the graph of Fig. 4, throughout all generations, the average payoff of unilateral choice continued to fluctuate near 0.6. The average payoff of mutual choice also continued to be near 0.7. The average payoffs of mutual choice with the split rule continued to slightly rise from 0.7 to 0.8. The average payoffs of mutual choice with the split rule and re-offering rose to 0.9 until 2,000 generations. After 2,000 generations, it seldom changed.

4   Discussion

We discuss the establishment of cooperation based on the strategy (C_i and V_i) in four cases and the enhancement of systems' performance based on the average payoff of all players. In the following discussion, a player with a high level of cooperativeness is represented as C_{high}, and a player with a low level of cooperativeness as C_{low}. In the same way, a player with high and low levels of vengefulness are represented as V_{high} and V_{low}, respectively.

4.1 Establishment of Cooperation

Case 1. In all trials in case 1, why did the establishment of cooperation fail? A player with low cooperativeness as the result of a mutation (i.e., a player with C_{high}), cannot acquire a higher payoff than the players with C_{low}. The reason is that the players with C_{high} cannot refuse an offer from the players with C_{low} in a group dynamic based on unilateral choice, so the players with C_{low} have a free-ride on the players with C_{high}. Accordingly, the players with C_{high} do not increase in the next generation. Therefore, cooperation is never established.

Case 2. In 36 trials of case 2, why did the establishment of cooperation fail? We analyze the factors leading to the failure of cooperation based mutual choice.

First, we consider the case where there is only one player with C_{high} and V_{high}. Mutation decreases cooperativeness or increases vengefulness because the initial condition is $V_i = 0$ and $C_i = 0$ ($\forall i \in N$). A player with cooperativeness decreased by mutation, i.e., a player with C_{high}, cannot acquire a higher payoff than the players with C_{low} because the players with C_{low} can freeride on the players with C_{high}. Accordingly, the players with C_{high} do not increase in the next generation. A player with vengefulness increased by mutation, i.e., a player with V_{high}, cannot acquire a higher payoff than the players with C_{low} because players with V_{high} do not join a group consisting of players with C_{low}. Therefore, one player with C_{high} and V_{high} by mutation cannot acquire a higher payoff than the players with C_{low} and V_{low}. Consequently, the player with C_{high} and V_{high} is not selected in the genetic operation, and so it perishes.

Next, we consider the case where there are several players with C_{high} and V_{high}. If a group consists of only players with C_{high} and V_{high}, the group refuses the offers of players with C_{low}. If a group consists of both players with C_{high} and V_{high} and players with C_{high} and V_{low}, it is possible that a player with C_{low} would join this group and be able to free-ride. The player with C_{low} can join the group because, while the players with C_{high} and V_{high} oppose the acceptance of his/her game offer, the players with C_{high} and V_{low} agree to it. If the players with C_{high} and V_{low} win the majority vote over the players with C_{high} and V_{high}, the player with C_{low} can join the group. In such a group, the players with C_{low} free-ride on the players with C_{high}. The players with C_{high} and V_{high} are not selected in the genetic operation and then perish because they cannot acquire higher payoffs than the free-rider. Although there are plural players with C_{high} and V_{high}, the players with C_{low} and the players with C_{high} and V_{low} prevent the establishment of cooperation. The players with C_{low} directly prevent the establishment of cooperation because they free-ride on the players with C_{high} and V_{high}. On the other hand, the players with C_{high} and V_{low} indirectly prevent the establishment of cooperation because they accept offers from the players with C_{low} who free-ride on the players with C_{high}. In the group dynamics based on mutual choice, therefore, the establishment of cooperation often fails.

On the other hand, in the remaining 4 trials, why did the establishment of cooperation succeed? Here, we analyze the factors leading to the establishment of cooperation in these cases. The reason for the establishment of cooperation was that a player with C_{low} can join the group consisting of both players with C_{high} and V_{high} and players with C_{high} and V_{low}. If there are players with C_{high} and V_{high} but no player with C_{high} and V_{low}, the player with C_{low} cannot join the group, and then the players will defect from each other. As a result, the player with C_{low} acquires a lower payoff than the players with C_{high} and V_{high} who cooperate with each other. If the number of players with C_{high} and V_{high} increases, and the players predominate in the population for a few generations before the number of players with C_{high} and V_{low} increases by crossover or mutation, cooperation becomes established. Therefore, since the simulation results show that cooperation was established in 4 out of 40 trials, we can conclude that it is not impossible but difficult to realize the establishment of cooperation in the group dynamics based on mutual choice.

Case 3. In 12 trials of group dynamics in case 3, why did the establishment of cooperation succeed? In case 3, the establishment of cooperation fails because the players with C_{high} and V_{low} accept the offer of the players with C_{low}. Here, we analyze the factors leading to the establishment of cooperation in the group dynamics based on mutual choice with the split rule.

In these group dynamics, if the players with C_{high} and V_{low} agree to accept the offer of a player with C_{low}, and the group as a whole also accepts it, the players with C_{high} and V_{high} leave the group based on the split rule; they refuse to play the dilemma game with those who play with defectors. The split rule prevents the player with C_{low} from having a free-ride on the players with C_{high} and V_{high}. This is because if the player with C_{low} joins the group, the players with C_{high} and V_{high} leave. As a result, if there are some players with C_{high} and V_{high}, they can form a group without the player with C_{low}. The players with C_{high} and V_{high} can acquire higher payoffs because they cooperate with each other. Throughout this process, the number of players with C_{high} and V_{high} increases and they predominate in the population. Therefore, cooperation becomes established.

Case 4. In all trials of group dynamics in case 4, why did the establishment of cooperation succeed? Based on group dynamics based on mutual choice and group split, a player with C_{high} and V_{high} sometimes leaves its group and then joins a group consisting of only itself. In this case, the player with C_{high} and V_{high} acquires the payoff of the loner, which is lower than those in a group. If the player with C_{high} and V_{high} has a chance of re-offering, the player leaving from one group may be able to join another group. The player with C_{high} and V_{high} can avoid acquiring a lower payoff by mutual cooperation if another group consisting of many players with C_{high} exists. The re-offering of a player leaving a group increases the chance for players with C_{high} and V_{high} to acquire a higher payoff. Accordingly, the establishment of cooperation increases because the players with C_{high} and V_{high} do not decrease in the next generation.

4.2   Comparison of Average Payoffs

In this research, we compare the effect of the proposed group dynamics using the average payoff because the average payoff can be considered as a measure of the system performance. Based on the comparison of the average payoffs, the effect of four cases is ranked in descending order as case 4, 3, 2, 1. >From the development of the average payoffs, we can acquire the following results concerning these group dynamics. In relatively early generations, the effect of the split rule doesn't provide good results for the establishment of cooperation because there is not a great difference between case 3 and 4.

5   Conclusions

In this paper, certain gourp dynamics were proposed in order to enhance the performance of systems of self-interested agents. The partner choice mechanisms for the multiple 2-PD were extended to that for a multiple N-person dilemma game to study this. Four kinds of group dynamics based on partner choice mechanisms were investigated: case 1) only group formation based on unilateral choice, case 2) only group formation based on mutual choice, case 3) group formation based on mutual choice and group split, and case 4) group formation based on mutual choice, group split and re-offering. In order to measure the effect of these on the establishment of cooperation and the enhancement of system performance, an agent-based simulation was used. Evolutionary agent-based simulations were conducted to confirm whether these group dynamics with the split rule could promote cooperative behavior of players and enhance the performance of systems.

On the establishment of cooperation, the following results were confirmed: in group dynamics with group formation based on only unilateral choice, it is impossible to establish cooperation. In group dynamics with group formation based on monly utual choice, it is not impossible but difficult to establish cooperation. Similarly, in group dynamics with group formation based on mutual choice and group dynamics, it is difficult to establish cooperation. In group dynamics with group formation based on mutual choice and group formation and re-offering, it is possible to reliably establish cooperation. Finally, it was confirmed that these group dynamics has a large enough effect to increase the performance of systems if these included group split and re-offering.

References

1. Ashlock, D., Smucker, S. and Stanley, A., Tesfatsion, L.: Preferential Partner Selection in an Evolutionary Study of the Prisoner's Dilemma. BioSystems **37** No. 1-2 (1996) 99–125
2. Axelrod, R.: An Evolutionary Approach to Norms. American Political Science Review **80** (1986) 1095–1111
3. Axelrod, R.: The Evolution of Cooperation. Basic Books, New York (1984)
4. Axelrod, R.: The Complexity of Cooperation. Princeton University Press, New York(1997)
5. Axtell, R.: The Emergence of Firms in a Population of Agents: Local Increasing Returns, Unstable Nash Equilibria, and Power Law Size Distributions. The Brookings Institution CSED Working Paper **3** (2000)
6. Axtell, R.: Non-Cooperative Dynamics of Multi-Agent Teams. Proceedings of The First International Joint Conference on Autonomous Agents and Multiagent Systems (2002) 1082–1089
7. Batali, J. and Kitcher, P.: Evolution of Altruism in Optional and Compulsory Games. Journal of Theoretical Biology **175** (1995) 161–171
8. Bicchieri, C.: Rationality and Coordination. Cambridge University Press (1993)

9. Caillou, P., Aknine, S. and Pinson, S.: A Multi-Agent Method for Forming and Dynamic Restructuring of Pareto Optimal Coalitions. Proceedings of The First International Joint Conference on Autonomous Agents and Multiagent Systems (2002) 1074–1081

10. Cohen, M. D., Riolo, R. L., Axelrod, R.: The Emergence of Social Organization in the Prisoner's Dilemma: How Context-Preservation and Other Factors Promote Cooperation. Santa Fe Institute Working Paper, 99-01-002, Santa Fe Institute (1999)

11. Dawes, R. M.: Social Dilemmas. Annual Review of Psychology **31** (1981) 169–193

12. Hauert, C., Monte, S., Hofbauer, J., Sigmund, K.: Volunteering as Red Queen Mechanism for Cooperation in Public Goods Game. Science **296** (2002) 1129–1132

13. Hauk, E. and Nagel, R.: Choice of Partners in Multiple Prisoner's Two-person Prisoner's Dilemma Games: An Experimental Study. Economics Working Papers, Universitat Pompeu Fabra (2000)

14. Hirshleifer, D. and Rasmusen, E.: Cooperation in a Repeated Prisoners' Dilemma with Ostracism. Journal of Economic Behavior and Organization **12** 87–106

15. Luis, J. and Silva, T.: Vowels Co-ordination Model. Proceedings of The First International Joint Conference on Autonomous Agents and Multiagent Systems (2002) 1129–1136

16. Ostrom, E.: Governing the Commons. Cambridge University Press, New York (1990)

17. Shussler, R.: Exit Threats and Cooperation under Anonymity. Journal of Conflict Resolution **33** (1989) 728–749

18. Stanley, E. A., Ashlock, D. and Tesfatsion, L.: Iterated Prisoner's Dilemma with Choice and Refusal of Partners. Artificial Life **III** (1994) 131–175

19. Soh, L. and Tsatsoulis, C.: Satisficing Coalition Formation Agents. Proceedings of The First International Joint Conference on Autonomous Agents and Multiagent Systems (2002) 1062–1063

20. Tesfatsion, L.: A Trade Network Game with Endogenous Partner Selection. Kluwer Academic Publishers (1997) 249–269

21. Yamashita, T. and Ohuchi, A.: Analysis of Norms Game with Mutual Choice. In Exploring New Frontiers on Artificial Intelligence, Springer-Verlag Tokyo (2002) 174–184

ASAP: Agent-Based Simulator for Amusement Park
— Toward Eluding Social Congestions Through Ubiquitous Scheduling —

Kazuo Miyashita

National Institute of Advanced Industrial Science and Technology,
Tsukuba, Ibaraki, Japan 305-8564
k.miyashita@aist.go.jp

Abstract. In this paper, an innovative application of scheduling methodology is advocated for the emerging service, which is named "social coordination" in the ubiquitous information environments. A typical service expected in ubiquitous computing is information provision adapted to each user's current situation. The service is supposed to increase a single person's convenience. However, a new type of service ("social coordination") is also possible for improving conveniences of the people sharing the ubiquitous information environment. The author explains the concept of "ubiquitous scheduling" that eludes congestions in the society by scheduling people's activities efficiently and rationally. To evaluate effectiveness of the concept, a multi-agent scheduler for an amusement park problem is implemented, which coordinates the demands for rides by tens of thousands people and makes suggestions as to when they should visit attractions in the amusement park to avoid standing in long lines.

1   Introduction

Due to drastic advancement of computers, sensors and wireless communication devices, we are entering a new era of ubiquitous computation [14]. However, in spite of the rapid progress of hardware technologies in ubiquitous computing, we are still in search of *killer applications* that take the most advantage of ubiquitous computation and give remarkable benefits to our daily activities.

So far, most of the applications proposed for the ubiquitous computing environment are to provide *context-aware information services* [10] for an individual user. For example, in a museum, based on a visitor's current location information, he/she can get an explanation on nearby paintings automatically via his/her mobile device (such as PDA or cellular phone). Moreover, based on information stored in the visitor's device, additional services (e.g., showing a way to pictures of a similar taste) can be provided for his/her convenience. This type of service is quite useful for each individual, but the service does not give any consideration to the relationship among people in their proximity. For example, the above system might cause congestion in the museum by guiding many people to the same picture in a short period of time.

In addition to the personally adapted information provision service proposed so far, a new type of service can be developed in the ubiquitous computing environment for

P. Davidsson et al. (Eds.): MABS 2004, LNAI 3415, pp. 195–209, 2005.
© Springer-Verlag Berlin Heidelberg 2005

eluding congestion in daily life. This type of service is called *social coordination* [4], since it pursues to coordinate people's behavior in the society for their conveniences and also contribute to improve social welfare.

In this paper, *ubiquitous scheduling* is advocated as one example of such social coordination services for reducing wasted time in congestions of the society. In ubiquitous scheduling, the social coordination problem is formulated as a multi-agent resource allocation problem and solved in real time using constraint-based scheduling heuristics. To validate effectiveness of the approach experimentally, the ubiquitous scheduling method is applied to the problem of making reservations of attractions for visitors in an imaginary amusement park. An amusement park simulator is implemented as a testbed and the simulation results are analyzed to see whether ubiquitous scheduling is useful for reducing congestions in the amusement park.

The paper's outline is as follows: Section 2 describes characteristics of the amusement park problem. Section 3 explains the detailed algorithm of ubiquitous scheduling. Then, in Section 4, the agent model and architecture of the implemented amusement park simulator is explained, and preliminary experimental results are shown and discussed in Section 5. The paper concludes with a summary and some suggestions for future work.

2   Amusement Park Problem

In an amusement park, we sometimes see popular attractions having long queues and, at the same time, some attractions running with few customers. In this paper, the author sets up an imaginary amusement park, which is microcosm of ubiquitous computing society, as a testbed for the study of social coordination. In the amusement park, visitors interact with the attractions and coordinate their plans of riding attractions to reduce congestion in the park. Recently the amusement park problem has attracted the attentions of the researchers [8, 3], because it has several practical and expedient reasons for investigation as follows:

– In an amusement park, visitors sometimes need to stand in line for an hour or more before they ride the most popular attractions in the park. And, to ensure repeat customers, amusement parks are interested in reducing the waiting hours for visitors and increasing their satisfaction. For that purpose, they have developed several systems such as wait hour display, but they are still in need of more effective solutions.
– Because the purpose of customers visiting an amusement park is to enjoy or ride as many attractions as possible during their visit, it does not contradict with the objective of the amusement park (i.e., to increase customers' satisfaction). Thus, benefits obtained by social coordination should be appreciated in the both sides.
– An amusement park is usually geographically isolated from the outside world. Therefore, it is plausible to reify ubiquitous computing environment for it.

2.1   Social Coordination in an Amusement Park

In the amusement park with ubiquitous computing capabilities, visitors are assumed to possess some portable terminal through which they can interact with attractions in the park and exchange information for the purpose of social coordination. Information to be transmitted from a visitor includes: time of admission, planned time to exit, attractions of interest and time range in which s/he wants to ride each attraction. Receiving information from the visitors, an attraction in the park aggregates its envisioned loads and suggests visitors when they should come for a ride. When the visitors accept the suggestion, the attraction makes a reservation of its capacity for their rides.

In a realistic situation, some visitors might dislike to use this system or disobey the suggestions made by attractions. Hence, it is important to give them incentives[1] for appreciating benefits of social coordination. In addition to such human problems on the user's side, the amusement park problem has the following computational difficulties:

Scale of problem: Millions of customers visit a large-scale amusement park in a year[2]. Hence, a large park usually has tens or hundreds of thousands of visitors in a day. And it has more than dozens of attractions. To coordinate the visitors' plans and produce reservations of *satisficing* quality for them, countless numbers of interactions among visitors and attractions might be required in the search of solutions.

Quick response: Information of visitors is not available prior to their visit to an amusement park. Therefore, social coordination cannot begin until they have arrived. But, it should not keep visitors waiting for the reservations to be made, since its objective is to reduce their waiting time. Social coordination should be executed in a timely manner.

Fairness of solution: For social coordination, another important factor to be considered is that the produced solution should appear fair to the visitors. Otherwise, the visitors will not follow the plans. Necessary information must be provided to visitors to make them believe in the fairness of the plan.

Although not all of the above problems are solved in the paper, the author proposes *ubiquitous scheduling* as a promising method for realizing a social coordination service. The ubiquitous scheduling method utilizes explicit representations of constraints in the problem and constraint-based heuristics for scheduling visitors' attraction rides. The following section shows how the ubiquitous scheduling method can solve the above problems by (1) reducing search space, (2) constructing problem space in an iterative and real-time fashion, and (3) providing meaningful problem solving information to visitors.

3   Ubiquitous Scheduling

In general, scheduling proceeds through interactions among two types of agents: (1) the user agent which represents a user who has temporal demands on some resources, and (2) the resource agent that manages capacity of a resource. The problem is to allocate capacity of resources to users, thereby satisfying users' demands as much as possible. To solve the problem in a distributed manner, several methods have been studied (e.g., contract-net [11], market oriented methods [7], biologically inspired approaches [1]). Those methods are either too time consuming or too ad hoc to solve a real-world problem.

---

[1] For example, each attraction gives higher priority to entry of visitors with reservation than those without reservation.

[2] Tokyo Disney Resort, which is one of the most popular amusement parks in the world, reported that about 24 million customers visited the park in 2003.

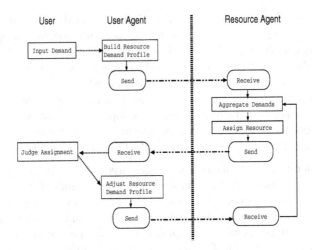

Fig. 1. Interactions among agents in ubiquitous scheduling

In the ubiquitous scheduling method, user agents and resource agents exchange their constraint information to search for a resource allocation solution that satisfies as many constraints as possible. Resource allocations found in the search are assigned to users as reservations of resources at the time interval. During the process, user agents do not interact with each other, neither do the resource agents. This kind of problem has been modeled and solved as a distributed constraint-based scheduling problem [12, 5]. For real-time problem solving, the ubiquitous scheduling method extends those past research results with a mechanism to relax constraints by promoting users' concession in necessary situations.

Fig. 1 depicts the interactions among agents in ubiquitous scheduling. Although user, user agent, and resource agent are shown as a single entity in the figure, in the implementation of ubiquitous scheduling, many users, user agents and resource agents interact with each other simultaneously and search for a feasible schedule iteratively. In the following subsections, details of the ubiquitous scheduling method are explained along the interaction flow shown in Fig. 1.

3.1   Demand Profile

When a user enters his/her demand information (i.e., earliest entry time and latest exit time for each resource in demand) through a portable terminal, the user agent, which is resident in the terminal, generates data called a *demand profile* [9] based on the input information. The demand profile represents a distribution of the user's demand strength on a resource along the timeline, considering the temporal constraints on the earliest entry and the latest exit.

Fig. 2 illustrates an example of a demand profile for the case where a user requests a resource in the time window between 10:00 and 10:50, and the resource needs 30 minutes for processing the user's request. In Fig 2 planning granularity is assumed 10 minutes. A smaller planning granularity produces a finer solution with more computation.

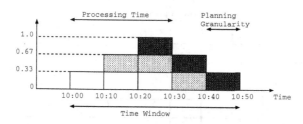

Fig. 2. Illustrated example of resource demand profile

Possible solutions for the user to satisfy his/her request are: to allocate the resource either (1) from 10:00 to 10:30, (2) from 10:10 to 10:40 or (3) from 10:20 to 10:50. Hence, distribution of user's demand strength on the resource takes a stepwise shape as shown in Fig. 2. The figure shows that, to satisfy user's request in this example, the resource must be allocated to the user from 10:20 to 10:30. In general, with a small time window, a demand profile takes large values near 1.0, which means that there is little choice of time for user's demand on a resource.

Before a user agent transmits a demand profile to a resource agent, the user's demand profile is normalized so that total demand to all the resources by the user should be equal to that of the other users. And, a demand profile needs to be re-calculated every time that the user's temporal constraints or resource's capacity constraints change in the process of ubiquitous scheduling.

3.2 Resource Allocation

After the user agents send the demand profiles to the corresponding resources, each resource agent aggregates the demand profiles that it receives. The aggregated demand profile shows *demand contentions* of users for the resource. For satisfying the users' demands as much as possible, the resource agent needs to search for allocation that resolves the contentions. Backtrack-based search used in the distributed constraint satisfaction methods [6, 15] is unsuitable for application to ubiquitous scheduling because it requires intolerable amounts of computation for solving a large scale problem. To avoid backtracks in the constraint satisfaction search process, the ubiquitous scheduling method uses constraint-based heuristics for making a partial solution. For the unsolved parts of the problem, the users are requested to modify or cancel their demands on the resource.

Fig. 3 depicts the demand profiles of 3 users and their aggregated demand profile of the resource. The graph of the aggregated demand profile (i.e., the bottom graph in Fig. 3) shows Resource1 has a *contention peak*. The demand profile graphs of 3 users reveal that User3 is the *most critical user* who has the largest demand at the contention peak of Resource1. Resource1 starts resolving contentions heuristically by allocating its capacity around the contention peak to User3 for his/her reservation, and then allocate the capacity away from the contention peak to the other users.

Different from the conventional reservation methods, which basically make reservations on the first-come-first-serve basis, the ubiquitous scheduling method considers tightness of users' constraints (i.e., demand contentions and length of demand time win-

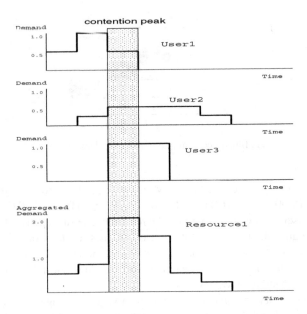

Fig. 3. Contention peak for resource

dow specified by a user) and gives higher priority to those with tighter constraints. But, to realize fair resource allocation among users, the ubiquitous scheduling method needs to force users to declare the tightness of their demand honestly. For the purpose, a resource agent modifies the demand profile from a user as follows:

$$d_u^* = \begin{cases} d_u w_u / T_r & \text{(if } w_u \le T_r) \\ d_u & \text{(otherwise)} \end{cases}$$

where d_u^* is a modified value of user's demand profile at a resource, d_u is the original value of the demand profile of the user on the resource, w_u is a time window length of the user's demand profile, and T_r is a standard time length defined by the resource. By this modification of the demand profile value, a resource can prevent users from setting a small time window to make their demand tight.

3.3   Demand Adjustment

When each resource agent makes decisions on allocation of its capacity to the users, it sends information about the allocation to the user agents.

Receiving the information, the user evaluates acceptability of the allocation as a reserved time slot for the resource. Because resource agents neither interact with other resource agents in allocating their own capacity nor have information about users' other fixed reservations, allocations made by the resource agents for a user might cause conflicts in his/her constraints. In distributed constraint satisfaction algorithms, those conflicts should be resolved by backtracking. However, in ubiquitous scheduling, the exponential explosion of search time caused by backtracking is not permissible. Therefore,

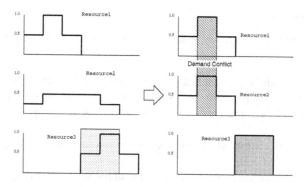

Fig. 4. Demand conflict for user

in ubiquitous scheduling, when a user has conflicts in his/her constraints, the user is required to adjust the constraints to resolve the conflicts. To help users make adjustments, they are provided information on constraints and conflicts.

Fig. 4 shows an example in which a user suffers a demand conflict from a newly assigned resource allocation. The left side of Fig. 4 depicts demand profiles of a user for 3 resources. The graphs on the right represent the modified demand profiles for the resources after the user accepts the assignment by Resource3 (i.e., a dotted area in the left graph). Since the user has to use Resource3 in the assigned time interval, the time window allotted for Resource2 should be shrunk. Hence, the demand profile for Resource2 changes its shape as shown in the figure. As a consequence, demand profiles of both Resource1 and Resource2 have a peak with a value of 1.0 at the same time interval. This situation is called a *demand conflict*. When this occurs, the user must decide whether (1) to reject an assignment by Resource3, (2) to accept the assignment by Resource3 and adjust his/her demands by giving up either Resource1 or Resource2, or (3) to give up Resource3.

Since rejection of the assignment by Resource3 causes similar extra computation with that of backtrack search, the maximum number of rejections allowed for a user should be limited. In ubiquitous scheduling, its effectiveness depends heavily on the ability to assign as many reservations as possible in a short time period. Thus, by putting a low limit on the number of rejections allowed, ubiquitous scheduling urges the users not to reject the assignments but to give up some of their demands. In other words, *mutual concession* by demand adjustment is requested of the users of ubiquitous scheduling.

4   Simulation System

The ubiquitous scheduling method is experimentally evaluated in the context of multi-agent simulation [2]. The ASAP (Agent-based Simulator for Amusement Park) system has been developed to test the effectiveness of ubiquitous scheduling as a social coordination service in realistic situations at an amusement park.

4.1   Agent Model

In the ASAP system, an imaginary amusement park consists of the following 3 types of agents:

1. **Visitor agent:** this is an agent that simulates behaviors of a customer who visits the amusement park for pleasure. In ASAP, the visitor agent does not have any direct interaction with the other visitor agents. Two types of visitor agents are assumed in ASAP;

    (a) *Cooperative visitors* who follow the reservation suggested by ubiquitous scheduling.

    (b) *Uncooperative visitors* who do not use ubiquitous scheduling and behave as they plan based on their individual priorities.

    Attributes of a visitor agent are: (1) time of entry, (2) planned time of exit, (3) walking speed, and (4) a list of attractions and their priority (and preferred time range of each visit as option).

    The cooperative agent v has an additional attribute α_v to be used when deciding whether it should go to the next reserved attraction, or stop by and stand in line for another attraction without reservation. In such a case, the agent who queues for the unreserved attraction may need to leave for another reserved attraction in time for its reservation. Probability p that the agent stops by at the attraction r^* before visiting the reserved attraction r is calculated as

$$p = 1.0 - \alpha_v w_{r^*}/s_r,$$

where w_{r^*} is waiting time for the attraction r^*, s_r is the time left till the reservation time of the attraction r. Hence, α_v shows cautiousness of the cooperative visitor agent v in deciding to visit the unreserved attractions.

The visitor agent can take one of the following states at a time: (1) idling, (2) walking to an attraction, (3) waiting without reservation, (4) queuing with reservation, (5) riding an attraction and (6) exit the park. Possible state transition for the visitor agent is shown in Fig. 5.

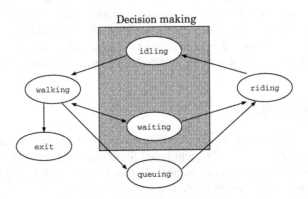

Fig. 5. State transition of visitor agent

The visitor agent make its decision on its behavior only in the idling state or in the waiting state. In the idling state, a visitor agent decides which attraction to visit next, and in the waiting state, a visitor agent decides whether to keep standing in line or leave for another attraction. After riding all the attractions on its list, the visitor agent walks to the exit and its state transits to the exit state.

2. **Attraction agent:** this agent simulates behaviors of attractions or other facilities in an amusement park such as restaurants or gift shops. As a consequence of behaviors made by the attraction agent, the states of visitor agents are changed accordingly. In ASAP, 2 types of attraction agents are assumed:

    (a) *Reservable attraction*: cooperative visitor agents can make a reservation with this type of attraction agent through ubiquitous scheduling. Uncooperative visitor agents also ride this type of attractions without a reservation, but this agent prioritizes cooperative visitors over uncooperative visitors. Hence, uncooperative visitors can ride the attraction only when the attraction has some capacity after satisfying all the reserved demands of cooperative visitors.

    (b) *Non-reservable attraction*: no visitor can make a reservation for this type of attraction. Hence, both cooperative visitors and uncooperative visitors should stand in the same line before riding the attraction.

3. **Coordinator agent:** this agent is in charge of making reservations for a reservable attraction through the ubiquitous scheduling method explained in Section 3. The coordinator agent does not interact with the other coordinator agents. In other words, it does not consider possible conflicts among reservations for a visitor. The conflicts should be resolved by the visitor agent.

As one of the current limitations of the ASAP system, it does not have models of "roads" in the park. It assumes that visitors can move from one attraction to another in a direct straight path with a constant walking speed. Another possible extension to the current implementation is to enable the visitor agents to change their attributes dynamically. This increase flexibility of the visitor's behavior, such as riding the same attraction again if the visitor might like it at the first ride.

4.2   Implementation

Fig. 6 shows an overview of the implemented architecture of the ASAP system. In the amusement park problem solved by ASAP, tens of thousands of visitors and dozens of attractions need to be simulated. Since it requires a considerable amount of computation, the ASAP system is implemented on the Beowulf cluster computer, which has 8 nodes of 3.06 GHz dual Xeon computers interconnected with Gigabit Ethernet. For communication among agents, distributed shared memory (i.e., tag boards in Fig. 6) is implemented to reduce the burden of programming tangled peer-to-peer interactions among many agents [13].

After a simulation run, ASAP can analyze the results graphically and quantitatively. With a graphical output shown in Fig. 7, ASAP can show the results of two simulation runs simultaneously as graphic animation, thus enabling the researcher to compare the dynamic aspects of the simulation results visually. For a more detailed and quantitative analysis, ASAP can also produce graphs of simulation results such as the utilization of attractions and the waiting time of visitors.

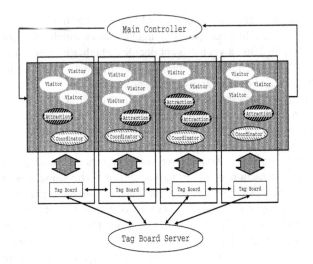

Fig. 6. Architecture of ASAP system

Fig. 7. Screenshot of ASAP system

5   Preliminary Experimental Results

To evaluate the effectiveness of the ubiquitous scheduling method, preliminary experiments of applying the ASAP system to an imaginary amusement park are executed.

5.1   Problem Definitions

An imaginary amusement park is defined for the experiments. The park is assumed to have 8 attractions (from Attraction1 to Attraction8, respectively) with the same capacity: they accommodate 130 customers for a ride at the maximum, take 5 minutes for the ride, and are open from 9 o'clock to 17 o'clock. 12,000 visitors enter the park at 9 o'clock.

They leave the park as soon as they finish riding all the 8 attractions in the park or when the park is closed at 17 o'clock.

The theoretical load ratio of every attraction in the park is 96%, which is calculated as follows:

$$L = V/C$$
$$= 12,000/(C_r N_r)$$
$$= 12,000/(130T/T_r)$$
$$= 12,000/(130 * (60 * (17 - 9))/5) = 0.96$$

where L is the theoretical load ratio of an attraction, V is the number of visitors to ride the attraction, C is the total capacity of the attraction in a day, C_r is the capacity of the attraction for a ride, N_r is the number of the rides operated by the attraction in a day, T is the total working hour of the attraction in a day, and T_r is the time needed for a ride. This load ratio (96%) means that the park has just enough capacity to satisfy all the demands by the visitors when they act in a coordinated way.

The priority list of attractions for a visitor is determined to make unevenly distributed preferences over the attractions with the following conditions: (1) One third of the visitors (i.e., 4,000 visitors) are designed to have the highest priority in visiting Attraction1 and the second highest priority in visiting Attraction2, (2) another one third of the visitors like to visit Attraction2 most and secondarily Attraction1, and (3) the rest of visitors are supposed to prefer visiting the other attractions over Attraction1 or Attraction2. This makes Attraction1 and Attraction2 more likely to be congested than the other attractions.

To see the effects of ubiquitous scheduling, the following 3 sets of the experiments are executed; (1) No Res.: no attraction has the reservation function, (2) Res.1: only Attraction1 has the reservation function with the ubiquitous scheduling method, and (3) Res.1&2: both Attraction1 and Attraction2 have the reservation function. Both in Res.1 and Res.1&2 cases, every visitor is assumed *cooperative* and follows the suggestions of ubiquitous scheduling.

Since an increase of attractions to be reserved complicates the coordination process among visitors and requires enormous computation, only the bottleneck resources (i.e., most popular attractions, Attraction1 and Attraction2) are considered in the experiments of this paper.

5.2   Experimental Results

Table 1 shows the average distribution of visitors' states in the 3 sets of experiments. From the result of No Res. case, a visitor spends 75% of the time in standing in line for attractions. In Res.1 or Res.1&2 cases, where ubiquitous scheduling is adopted, time spent in waiting and queuing is about 65%. Thus, by making reservations for popular attractions (i.e., Attraction1 and Attraction2), a visitor can reduce his/her waiting time by about 10%.

The number of attractions ridden by a visitor in Fig. 8 shows drastic distinctions among the 3 cases. In the No Res. case, although 2,903 visitors can ride all the attractions, 2,243 visitors ride less than 4 attractions. But in the Res.1&2 case, almost all the visitors (11,705 visitors out of 12,000) ride all the attractions and the rest of visitors ride 7

Table 1. Visitor's state distribution

	No Res.	Res. 1	Res. 1&2
idling	1	1	1
walking	15	17	18
queuing	0	11	13
waiting	75	53	52
riding	6	8	8
exit	3	10	8

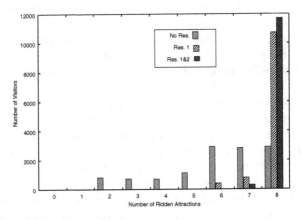

Fig. 8. Distribution of visitors with a number of attraction rides

attractions. This result suggests that ubiquitous scheduling can give fair and extended opportunities to all the visitors even when the attractions have a high theoretical load ratio of 96% and the visitors have unbalanced priorities to the attractions. Thus, it is shown that ubiquitous scheduling improves visitors' satisfaction with the park.

The above discussions explain the results on the visitor's side. For management of the amusement park, in addition to improving customer's satisfaction, keeping high utilization of attractions is also important since it costs a large amount of money to build an attraction [3] and it may become obsolete in a year or two. Fig. 9 depicts the average utilization of the attractions in 3 cases of experiments. In the No Res. case, although popular attractions have high utilization, the other attractions have a considerable loss of utilization. This is because visitors' time is wasted waiting in the lines of popular attractions. In the Res.1 and Res.1&2 cases, all of the attractions have high utilization of more than 90%.

Fig. 10 presents the average queue length for each attraction in 3 cases of the experiments. In the No Res. and Res.1 cases, long queues are formed at the popular attractions (i.e., Attraction1 and Attraction2). And, in the Res.1&2 case, the queues are evenly distributed among all the attractions, thus visitors do not need to wait more than 1 hour for

---

[3] Some attraction facilities might cost more than 10 million dollars.

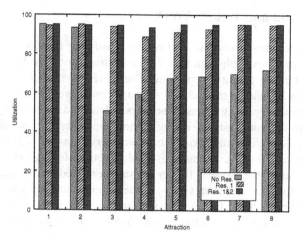

Fig. 9. Utilization of attractions

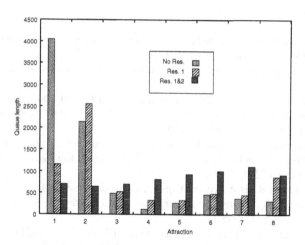

Fig. 10. Average queue length of attractions

any attraction. At the same time, all the attractions have enough length of visitor's queue to achieve high utilization.

These results show that making reservations only at the popular attractions can improve visitor's satisfaction at the amusement park. Although taking reservations at all the attractions can further reduce visitor's waiting time at the unpopular rides, it might not be worth the required computational cost when considering the expected extra benefits of the visitors. Giving an opportunity of making reservations at the popular attractions to the visitors is practically beneficial and feasible suggestion to the amusement parks.

6 Conclusion and Future Work

In this paper, the author advocates *ubiquitous scheduling* for eluding congestion in society. It shows an example of a *social coordination* service to be provided in the coming ubiquitous computing environments. The ASAP system is developed as a testbed of the ubiquitous scheduling method, and an imaginary amusement park is modeled and analyzed for evaluating effectiveness of the proposed method. Preliminary experimental results show that the ubiquitous scheduling method using constraint-based heuristics can solve a large scale problem of making attraction reservations for 12,000 visitors in the amusement park and produce fair and reasonable solutions for visitors.

As an extension of the research, further investigation is required to examine practical applicability of the method. For this purpose, the author is now applying the ubiquitous scheduling method to the data of a real amusement park in Japan. And, future work should also address the issue of efficient coordination of a visitor's reservations made by many attractions.

References

1. Eric Bonabeau, Marco Dorigo, and Guy Theraulaz, editors. *Swarm Intelligence: From Natural to Artificial Systems*. Oxford University Press, 1999.
2. J.M.Epstein and R.Axtell. *Growing Artificial Societies - Social Science from the Bottom UP*. MIT Press, 1996.
3. Hidenori Kawamura, Koichi Kurumatani, and Azuma Ohuchi. Modeling of theme park problem with multiagent for mass support. In *Proc. of The IJCAI-03 Workshop on Multiagent for Mass-User Support*, pages 1–10, 2003.
4. Koichi Kurumatani. Social coordination with architecture for ubiquitous agents: Consorts. In *Proc. of International Conference on Intelligent Agents, Web Technologies and Internet Commerce IAWTIC'2003*, 2003.
5. Jyi-Shane Liu and Katia P. Sycara. Exploiting problem structure for distributed constraint optimization. In *Proceedings of the First International Conference on Multi-Agent Systems*, pages 246–253. AAAI, 1995.
6. Kazuo Miyashita. CAMPS: A constraint-based architecture for multiagent planning and scheduling. *Journal of Intelligent Manufacturing*, 9:147–154, 1998.
7. M.P.Wellman and W.E.Walsh. Auction protocols for decentralized scheduling. *Games and Economic Behavior*, 35:271–303, 2001.
8. Jorge E. Prado and Peter R. Wurman. Non-cooperative plannning in multi-agent, resource-constrained environments with markets for reservations. In *AAAI whorkshop planning with and for Multiagent Systems Technical Report WS-02-12*, pages 60–66, 2002.
9. Norman Sadeh. Micro-opportunistic scheduling: The micro-boss factory scheduler. In M. Zweben and M. Fox, editors, *Intelligent Scheduling*. Morgan Kaufmann, San Mateo, CA, 1994.
10. B. Schilit, N. Adams, and R. Want. Context-aware computing applications. In *Proc. of IEEE Whorkshop on Mobile Computing Systems and Applications*, pages 85–90, 1994.
11. Reid G. Smith. The contract net protocol: High-level communication and control in a distributed problem solver. *IEEE Transactions on Computers*, C-29(12):1104–1113, 1980.
12. Katia P. Sycara, Steven F. Roth, Norman Sadeh, and Mark S. Fox. Resource allocation in distributed factory scheduling. *IEEE Expert*, 6(1):29–40, 1991.

13. Holger Veit and Gernot Richter. The FTA design paradigm for distributed systems. *Future Generation Computer Systems*, 16:727–740, 2000.
14. Mark Weiser. Hot topic: Ubiquitous computing. *IEEE Computer*, pages 71–72, 1993.
15. Makoto Yokoo, Edmund H. Durfee, Toru Ishida, and Kazuhiro Kuwabara. Distributed constraint satisfaction for formalizing distributed problem solving. In *Proceedings of the Twelfth International Conference on Distributed Computing Systems*, pages 614–621, 1992.

Patchiness and Prosociality: An Agent-Based Model of Plio/Pleistocene Hominid Food Sharing

L.S. Premo

Department of Anthropology, University of Arizona,
Tucson, Arizona, 85721, USA
lpremo@email.arizona.edu

Abstract. Anthropologists have yet to adequately investigate the evolution of food sharing despite its prevalence among contemporary human societies. As an initial step toward rectifying this lapse, I present preliminary population-genetic results generated by an agent-based model of Plio/Pleistocene hominid food sharing. SHARE explores the dynamics of a unique conceptual model that treats fragmented closed habitat patches as the loci of hominid social evolution and investigates the altruistic behavior of food sharing using multilevel selection theory. Data collected from artificial societies of hominid foragers demonstrate that specific levels of ecological patchiness facilitate the evolution of food sharing due to the fitness benefits bestowed upon subsistence–related trait groups.

1 Introduction

Multi-agent-based modeling provides researchers interested in the evolution of cooperation with a unique and sophisticated methodological tool (e.g., Hales, this volume). To wit, the research presented below employs artificial societies to investigate how food sharing, a biologically altruistic behavior displayed (to some degree) by all living humans, could have evolved[1] in ancestral populations and to explore the recursive roles that ecology and behavior might have played in this biosocial process during the period known as the Plio/Pleistocene (~3.5 million – 1 million years ago).

1.1 The Study of Altruistic Food Sharing and Hominid Evolution

Biologically altruistic behaviors benefit the fitness of one or more recipients while incurring a cost to that of the actor/donor. Examples include alarm calling, feeding restraint, and—the topic of this paper—food sharing. Food sharing is but one of many intriguing prosocial behaviors exhibited by living humans. Although ethologists recognize that ours is not the only species to display this altruistic behavior, both the

---

[1] Traditionally, "evolution" is defined as a *change* in the allele frequency of a population through time. However, for the purposes of this project, when used in conjunction with a particular allele, trait, or behavior, the terms "evolution," "evolve," "evolved," and "spread" synonymously refer to the *increase* in the frequency of that phenotype through time.

P. Davidsson et al. (Eds.): MABS 2004, LNAI 3415, pp. 210–224, 2005.

rich diversity of food types involved and the high frequency at which food transfers occur among non-related individuals distinguish our version as quantitatively, if not qualitatively, unique from those documented in other animal societies. Because the majority of anthropological research on human food sharing has focused on explaining its function in extant hunter-gatherer communities rather than on tracing its evolutionary history, important questions concerning when and how food sharing evolved in hominid populations have been left unaddressed.

In the absence of unequivocal archaeological evidence and analytic models, previous attempts to reconstruct Plio/Pleistocene hominid food sharing behaviors have relied heavily upon referential modeling, an inherently limiting enterprise that employs an observable phenomenon (contemporary hunter-gatherer behavior) as a model for a targeted referent phenomenon which may be impossible to study directly (hominid behavior) [18]. Many of the reconstructions yielded by this methodology were viewed as controversial because they casually ascribed contemporary human food sharing behaviors to hominids living nearly two million years ago [5,7]. In response, some anthropologists issued warnings against the unfettered use of living hunter-gatherer (and nonhuman primate) behaviors as referential models for interpreting early hominid material remains [18].

1.2 Agent-Based Modeling Provides a New Method to Investigate an Old Question

Though adept at inferring behavioral processes from scarce archaeological assemblages, paleoanthropologists yearn for an opportunity to observe their subjects firsthand. To some extent, this can be accomplished through conceptual models that afford the opportunity to observe cultural formation processes as they unfold in caricatures of past societies. Archaeologists often compare conceptual model output to empirical observations in order to learn more about the range of processes that *could have been* responsible for the patterns they exhume in the field. It is in this capacity—as tools used to explore one's ideas about the past—that models are most appropriately employed in archaeological research. Despite their long-standing relationship with other types of conceptual models, archaeologists have only recently become acquainted with agent-based simulations. Though the majority of the discipline views the first wave of agent-based applications with an appropriate mix of skepticism and curiosity, a distinguished minority has already begun to extol the methodology's exciting potential [6].

The agent-based model introduced here was built to address two principal questions, the first of which serves as the focus of this paper. Given certain assumptions and constraints, what range of ecological and social conditions facilitates the evolution of food sharing in artificial Plio/Pleistocene hominid populations? Second, does food sharing imprint a diagnostic spatial signature on archaeological landscapes, and, if so, how can this possibly multi-scale pattern be recognized in both experimental and empirical distributions of material culture? Previous studies relied heavily upon observations of living humans to address food sharing issues. Here, artificial societies of hominid foragers, which can be placed in any number of ecological scenarios and imbued with a variety of social rules, are employed as so-called cultural laboratories to investigate a much wider range of behavioral possibilities, many of which are not observable today.

2 A New Conceptual Model of Hominid Food Sharing

As we shall see in this section, previous anthropological thought on the topic of food sharing focused on selective pressures associated with open grasslands and searched for a selfish motivation for sharing in the form of benefits to the donor's fitness. As a response to these studies, my theoretical approach differs fundamentally in two respects. To more accurately characterize the socio-ecological milieu of Plio/Pleistocene hominids, I 1) focus on the selective pressures associated with fragmented patches of closed (woodland) habitat and 2) expand evolutionary ecological explanations of food sharing to include the selective benefits bestowed upon supra-individual vehicles of selection (i.e., trait groups). This multilevel selection perspective marks a significant departure from previous anthropological reconstructions of early hominid food sharing, and it yields a new conceptual model in which the altruistic trait evolves to fixation due to the benefits it bestows upon the fitness of subsistence-related trait groups competing with one another in a heterogeneous environment. This model implies that the strategy of sharing patchy woodland resources, not stalking prey in open grasslands, might have laid the ethological foundation for what we recognize today as exceptionally cooperative human societies.

2.1 East African Paleoecology

Raymond Dart's [2] colorful, but ultimately inaccurate, behavioral reconstruction of *Australopithecus africanus* as a formidable savanna carnivore is largely responsible for the popular notion that Plio/Pleistocene hominids were endemic to open grasslands. Seemingly bolstered by Vrba's [19,20] extensive research on climate change in sub-Saharan Africa, this deep-seated notion has been invoked by paleoanthropologists interested in such diverse topics as the emergence of bipedalism, the process of encephalization, early hunting/scavenging opportunities, and food sharing.

Today few paleoanthropologists would deny that a general cooling and drying trend allowed open grasslands to expand at the expense of closed forests and woodlands at various times during the Pliocene and early Pleistocene in sub-Saharan Africa. However, not every researcher feels that the role open grasslands played in hominid evolution was as direct and prominent as earlier explanations might lead one to believe. In fact, contrary to the predictions of the traditional savanna hypothesis, Plio/Pleistocene hominid skeletal and material remains have been found in association with a variety of paleohabitats including swamps, treeless to wooded grasslands, woodlands, and gallery forests [14]. In addition, postcranial morphological traits of *Australopithecus afarensis*, such as long curved phalanges and a relatively large humero-femoral index, serve as skeletal evidence that these hominids possessed physical adaptations for life in and around the trees [16,17].

The expansion of open grasslands during the Pliocene and early Pleistocene undoubtedly presented a significant environmental change that had far-reaching effects on the community ecology of sub-Saharan Africa. However, I believe the evolutionary significance of this ecological shift derived not from the *expansion of open habitat*, per se, but rather from the *fragmentation of closed habitat*. It is difficult to accept that late Pliocene hominids—seemingly still physically adapted to closed habitats—would have voluntarily abandoned habitable woodland patches in order to compete for the treeless ecological niche. The inverse of this popular explanation embodies a more plausible evolutionary scenario. That is, in the face of significant ecological changes caused by a cooling and drying climate, hominid populations continued to adapt (in this case, largely behaviorally) in ways that enhanced their utilization of woodland patches, not open grasslands. The loci of Plio/Pleistocene hominid social evolution, therefore, can be found in disjointed patches of closed habitat, not in treeless expanses.

2.2 Expanding Evolutionary Explanations of Food Sharing

Those who perform biologically altruistic acts, such as sharing food, benefit a recipient's individual fitness at a cost to their own. Because a selfish individual (social cheater) is able to enjoy the benefits of another's selfless deed without paying the associated costs, each egoist possesses a higher relative fitness than that of whom s/he takes advantage. How, then, can an altruistic trait evolve to fixation in a mixed population if individuals who display it are less fit than individuals who do not? In the past, attempts to make food sharing evolve via individual selection, thereby avoiding this thorny issue, have discredited the self-sacrificial nature of the behavior [1]. But recently strides have been made to couch explanations of food sharing in a multilevel selection evolutionary framework, one which allows researchers to more accurately model altruistic traits [21].

Although it requires that we expand our explanations of food sharing to include fitness benefits bestowed upon evolutionarily meaningful groups of individuals, multilevel selection theory is an elegant conceptual framework predicated on the idea that natural selection concurrently operates at two levels of the biological hierarchy: *within-group* and *between-group* [15,23]. Synonymous with individual selection, within-group selection promotes phenotypic traits that allow an individual to maximize one's relative fitness, regardless of how one's actions affect the fitness of others. On the other hand, between-group selection promotes phenotypes that are beneficial to the fitness of others while costly to that of the actor. The key to understanding this fundamental tenet of multilevel selection lies in the concept of trait groups. Trait groups are not defined spatially or by common descent (but note that spatially proximate individuals [10] and kin [4] can sometimes meet trait group requirements). Rather, a trait group is simply "a set of individuals that influence each other's fitness with respect to a certain trait but not the fitness of those outside the group" [15: p 92]. A population that is divided into trait groups will often experience increased levels of between-group selection which can foster the evolution of altruism, as we shall see below.

2.3 Previous Research on Patchiness and Prosociality

John Pepper and Barbara Smuts [9] present an elegant multilevel selection simulation, called ECO, which they use to study the evolution of two altruistic traits—alarm calling and feeding restraint—in freely-mixing populations of generalized foragers. They report that when inhabited by a mixed population of altruists and nonaltruists, patchy resource distributions can support the spread of these particular altruistic traits by *facilitating* (not to be confused with *causing*) positively assorted interactions among socially inept[2] foragers. In the absence of cultural mechanisms responsible for facilitating nonrandom interactions between similar individuals, positive assortment resulting from resource patchiness effectively structures mixed populations into evolutionarily meaningful trait groups.

Pepper and Smuts' conclusion that "[trait] groups emerging through the behavior of individual agents in patchy environments are sufficient to drive the evolution of group beneficial traits" [9: p 70] could prove crucial to an understanding of early hominid food sharing, given the patchy resource structure of Plio/Pleistocene savannas. As open grasslands encroached upon woodlands, resource patchiness would have facilitated assortative interactions within hominid subgroups relegated to slowly shrinking islands of preferred closed habitat. Pepper and Smuts' findings suggest that these fragmented environmental conditions could shift the balance of selective pressure from within-group to between-group, thereby fueling the spread of food sharing in the metapopulation through the differential reproductive success of trait groups. Testing this hypothesis requires a different agent-based model, one that explores dynamics particular to food sharing among abstracted hominid foragers.

3 SHARE: An Agent-Based Model of Hominid Food Sharing

Agent-based models allow social scientists to investigate emergent group level phenomena resulting from historically contingent interactions between heterogeneous agents, each of which behaves according to rule-based schemata [3]. Free from the deterministic, top-down structure of traditional equation-based simulation techniques, evolutionary agent-based models provide a generative, bottom-up understanding of selective processes as simulated societies evolve according to particularistic population dynamics. The following paragraphs briefly describe an agent-based model, aptly named SHARE (Simulated Hominid Altruism Research Environment), which was implemented in Objective-C using the Swarm libraries [8] in order to explore the population-genetic consequences of the conceptual model presented above. SHARE's world is a two-dimensional grid, wrapped into a torus to avoid edge effects. Three types of agents—plants, meat, and foragers—can occupy each regularly shaped, regularly spaced grid cell.

3.1 The Physical Environment: Food Resources

Plants. Plant agents represent closed habitat (woodlands) food resources. Therefore, areas that lack plant agents model open grasslands. To characterize open *grass*lands as void of plant foods might seem counterintuitive, but because paleoanthropologists do not widely consider edible grasses as important components of most early hominid diets, they are not emphasized in SHARE. At the start of each simulation, plant agents are systematically distributed into regularly shaped and regularly spaced patches according to two experimental variables, patch size and gap size (Fig. 1). Plant agents do not die, move, or reproduce. Each plant's energy store represents the amount of

---

[2] Unable to discriminate the identities, phenotypes, or relatedness of other foragers.

energy that a forager could potentially gain by consuming it. When not fed upon, a plant's energy store grows logistically up to a relatively low fixed maximum. To foragers, plant agents represent temporally and spatially reliable, albeit relatively small, packages of food.

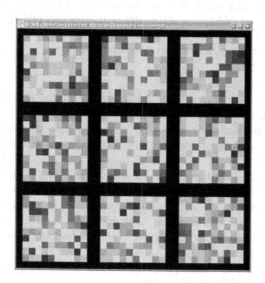

Fig. 1. Plant agents arranged in patches at the start of a SHARE simulation run (Patch Size = 10 and Gap Size = 2). Plant energy level is depicted in grayscale: higher values are lighter

Meat. Meat agents represent the carcasses of the medium to large-sized mammals often associated with both open and closed components of African savanna environments. Unlike plants, meat energy quickly depletes from a relatively high maximum value according to a logistic decay rate to emulate scavenging by other savanna carnivores and/or decomposition. Also unlike plant agents, meat agents are randomly instantiated over space and through time. The probability that a meat agent will be added to each cell per time step is an experimental variable that can be varied to investigate the impact that large, unpredictable food packages have on the evolution of food sharing. Though capable of yielding much more energy than a plant, each meat agent provides a spatially and temporally unreliable food source. Though meat agents are described here, note that they were not included in any of the simulation runs that yielded the results reported below.

3.2 The Social Environment: Hominid Foragers

Foraging, Reproduction, and Death. Forager agents are purposefully abstracted versions of Plio/Pleistocene hominids. Foragers that successfully procure plants and/or meat on a regular basis amass healthy stores of energy that eventually surpass their fertility thresholds. Forager reproduction is asexual and entails a significant energetic cost to the parent. To guarantee that each offspring inherits its parent's genotype, genetic transmission occurs through a single haploid locus with two immutable alleles (S = selfish and A = altruistic). Foragers that metabolize all of their stored energy while in search of food die of starvation. Foragers are also removed from the simulation if they fall prey to a savanna carnivore, disease, or an accident or when they reach their maximum age. The probability of dying from something other than old age or starvation is an experimental variable that can vary between open and closed habitats (the former more dangerous than the latter). Note that this particular experimental variable is not employed in the baseline version of SHARE.

Movement. Each forager first scans its immediate Moore neighborhood (including its current cell) for plant and meat agents. Of the subset of cells that contain enough food to support the metabolic cost of at least one time step and are not currently occupied by another forager, foragers will decide to move to that which contains the highest plant and/or meat energy level (ties broken randomly). If adequate food resources cannot be found in the adjacent cells, then the forager moves to one of them at random. While in the process of scanning the Moore neighborhood for food, each forager also looks for other foragers. If necessary, the first forager spotted in this manner can be asked to share excess food[3] during the current time step. These simple foraging rules effectively model hominids using individually retrieved spatial information to exploit locally available resources. The use of random walks between closed patches insures that all foragers employ the same strategy for finding additional woodland resources.

Food Sharing. To make food transfers possible, foragers are allowed to carry limited amounts of procured but unconsumed energy. Each forager possesses a (possibly unique) floating-point value greater than or equal to zero and less than or equal to one that provides the probability that it will share excess food when approached by another in need. Foragers are considered to be in need when their current energy level drops below the food share threshold value. In its current form SHARE has the capability to model three distinct sharing behaviors, described below in order of increasing social sophistication.

According to the simplest sharing rule (Rule 1), a donor (D) shares excess food with a prospective recipient (R) when a random number between zero and one is less than or equal to D's food share probability. The second food sharing method (Rule 2) resembles the well-known tit-for-tat strategy, and it requires that foragers remember whether agents cooperated with or defected against them in their most recent social interaction. Memory is operationalized with dynamic lists of social cooperators and defectors. Each agent maintains its own lists by updating them after each interaction in which it functioned as the prospective recipient. Donors use their memory to make food sharing decisions on a forager-by-forager basis according to the following rule: share excess food with those who shared with you in the most recent interaction, but refuse to share with those who refused to share with you. If D has no memory of a past interaction with R, Rule 1 is used instead. According to the

---

[3] The amount of excess food is defined as the positive difference of a forager's current energy level and the food share threshold value.

final food sharing method (Rule 3), D can accurately identify whether R is altruistic or selfish. D uses this important information to decide whether or not to share according to the following rule: altruistic D share with altruistic R but not with selfish R, and selfish D share with no one. Depending on one's theoretical proclivity, Rule 3 models either the situation in which there exists a reliable costly signal for cheaters, cooperators, or both or the scenario in which foragers use information exchanged with others (via gossip) about past social interactions to identify and act appropriately towards individuals upon their first meeting.

3.3 Variable Settings and Parameter Sweeps

To explore how a variety of socio-ecological contexts affect the evolution of the food sharing strategies described above, 30 simulation runs—each initialized using the same set of standard variable values (Table 1) and one of 30 unique random number seeds—were executed for each possible combination of experimental variables (e.g., Patch size 5, Gap size 8, Rule 1).

Table 1. Standard and experimental parameter settings for the baseline version of SHARE

Parameter	Value(s)
Standard Variables	
Minimum number of plants	500
Plant maximum (energy units)	10
Plant logistic growth rate r	0.2
Starting number of foragers	40 (20 of each type)
Forager starting energy (energy units)	50
Forager metabolic rate (energy units)	2
Forager fertility threshold (energy units)	100
Forager birth interval (time steps)	20
Cost of reproduction (energy units)	50
Forager maximum life span (time steps)	100
Forager maximum (energy units)	110
Forager food share threshold (energy units)	50
Probability of sharing food	0 (selfish), 1 (altruist)
Experimental Variables	
Patch size (number of cells per patch side)	1, 2, 3, 4, 5, 6, 7, 8, 9, 10
Gap size (number of cells between patches)	1, 2, 3, 4, 5, 6, 7, 8, 9, 10
Food sharing behaviors	Rule 1, Rule 2, Rule 3

3.4 Data Collection

Price's equation [11,12,13] provides an elegant tool for quantifying the relative strengths of between-group and within-group selective pressures as they fluctuate through time. By employing patch membership as a reliable proxy for trait group affiliation in SHARE, Price's equation can be used to partition the overall change in allele frequency so that selection within and between subsistence-related trait groups can be tracked separately. During each simulation run, the values yielded by

Price's equation, the percentage of altruistic foragers present in the metapopulation, and the total number of foragers in the metapopulation were collected to confirm that altruistic alleles increased in frequency due to raised levels of between-group selection and not by other means.

4 Results

I have qualified the results presented here as preliminary not to indicate that the data will change, but rather to underscore the fact that more will be added before completing the full exploration of this conceptual model's population-genetic and archaeological consequences. For instance, as mentioned above, this fundamental version of the model does not include large unpredictable energy packages (meat), mortality due to predation/disease/accident, or polygenic food sharing phenotypes—all of which are important aspects of the conceptual model that are currently being addressed with additional parameter sweeps—nor does it consider in any way the archaeological manifestations of food sharing. Thus, Figures 2-4 summarize only the population-genetic results of 9,000 runs of the most basic version of SHARE.

In considering Figures 2 and 3, the effect of ecological patchiness is readily discernible—altruistic food sharing evolves to fixation predominantly in intermediate patch sizes (3-6) when combined with intermediate-to-large gap sizes (3-10). This pattern echoes that which Pepper and Smuts [9] found for feeding restraint and alarm calling, but why? Recall that the evolution of altruism requires strong between-group selection. The strength of between-group selection depends largely upon the partitioning of genetic variation within and among trait groups. Resource patchiness can structure the genetic variation of a population in a variety of ways, some of which actually weaken between-group selection: small patches succeed in creating trait groups that are too ephemeral to compete as evolutionarily meaningful groups; large patches can support large trait groups that often contain a mix of forager types, thereby effectively increasing both within-group heterogeneity and between-group homogeneity; small gaps between patches do not pose deterrents for migration; and large gaps present serious obstacles that make it nearly impossible for reproductively successful trait groups to export their offspring to additional patches [22]. As Figures 2 and 3 clearly illustrate, between these extremes exists a transitional range of resource patchiness that provides the structure necessary to form internally homogenous and externally heterogeneous trait groups. It is under this rather restricted range of conditions that ecological patchiness enjoys its most influential between-group selective power. The results show that, though theoretically possible, it is far less likely for altruistic versions of Rules 1 and 2 to evolve to fixation in environmental settings outside of this so-called transitional range of resource patchiness. The fact that altruistic alleles evolve to fixation less frequently than their selfish counterparts should not detract from the finding that ecological heterogeneity can effectively structure forager populations into evolutionarily meaningful groups.

Gap Size

		1	2	3	4	5	6	7	8	9	10
	1 |	0	*	*	*	*	*	*	*	*	*
	2 |	0	1	*	*	*	*	*	*	*	*
	3 |	0	0	1	5	*	*	*	*	*	*
Patch	4 |	0	0	0	1	5	7	2	2	*	0
Size	5 |	0	0	0	2	0	0	3	9	9	6
	6 |	0	0	0	0	0	0	0	0	1	5
	7 |	0	0	0	0	0	0	0	0	0	0
	8 |	0	0	0	0	0	0	0	0	0	0
	9 |	0	0	0	0	0	0	0	0	0	0
	10 |	1	0	0	0	0	0	0	0	0	0

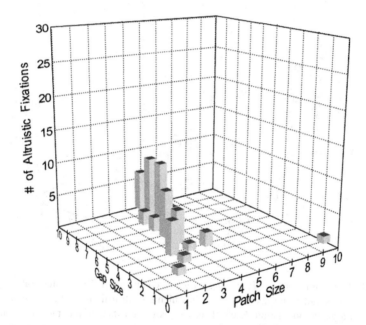

Fig. 2. Rule 1 results for each *Patch Size/Gap Size* combination. Table values (above) represent the number of simulation runs (out of 30) in which the altruistic food sharing allele evolved to fixation in a viable[4] population. Each of the values greater than zero is plotted in the bar chart along the axis labeled *# of Altruistic Fixations*. *None of the runs yielded a viable population at time of allele fixation

---

[4] In this study, "viable" populations contain at least ten foragers (25% of the starting population number). This arbitrary threshold effectively eliminates from further consideration most of the populations near extinction at the time of allele fixation.

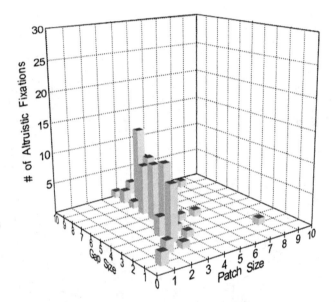

	Gap Size									
	1	2	3	4	5	6	7	8	9	10
1 |	2	*	*	*	*	*	*	*	*	*
2 |	0	2	0	*	*	*	*	*	*	*
3 |	0	1	9	3	*	*	*	*	*	*
Patch 4 |	0	0	1	1	10	9	8	1	2	1
Size 5 |	0	0	0	0	1	5	3	8	12	2
6 |	0	0	0	0	1	0	0	2	3	6
7 |	0	0	0	0	0	0	0	0	0	0
8 |	0	1	0	0	0	0	0	0	1	0
9 |	0	0	0	0	0	0	0	0	0	0
10 |	0	0	0	0	0	0	0	0	0	0

Fig. 3. Rule 2 results for each *Patch Size/Gap Size* combination. Table values (above) represent the number of simulation runs (out of 30) in which the altruistic food sharing allele evolved to fixation in a viable population. Each of the values greater than zero is plotted in the bar chart along the axis labeled *# of Altruistic Fixations*. *None of the runs yielded a viable population at time of allele fixation

Here, it is important to remember that the significance of an event need not be directly related to its probability, for rare events often precipitate serious consequences. In this sense, the quantitative results presented in this section can also be interpreted at a qualitative level (binary: presence/absence).

Some of the results obtained from this relatively conservative model are surprising. The fact that even the least sophisticated form of food sharing (Rule 1) occasionally evolved to fixation, despite the fact that it involves neither memory of past interactions nor the ability to identify the phenotype of prospective recipients, demonstrates the powerful role that resource patchiness can play in structuring an otherwise freely-mixing population of socially inept foragers. Less surprising, but still interesting, is the observation that the selective influence of ecological patchiness is inversely related to the social sophistication of food sharing behaviors. Figures 2 and 3 show that resource patchiness played an instrumental role in the evolution of the socially sophomoric food sharing behaviors (Rules 1 and 2), both of which evolved to fixation under a relatively restricted range of ecological conditions. In contrast, Figure 4 shows that resource patchiness played only a minor role in the case of the culture-laden Rule 3, which spread to fixation at least once under each and every ecological

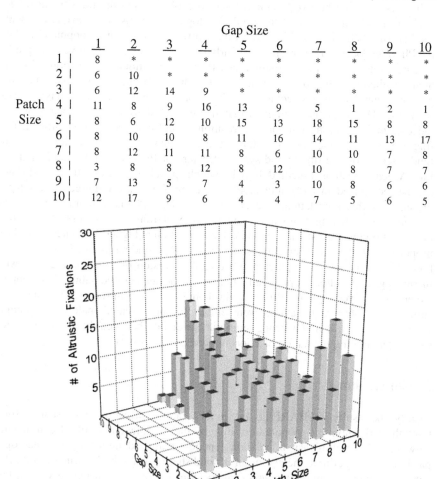

		Gap Size								
	1	2	3	4	5	6	7	8	9	10
1 |	8	*	*	*	*	*	*	*	*	*
2 |	6	10	*	*	*	*	*	*	*	*
3 |	6	12	14	9	*	*	*	*	*	*
Patch 4 |	11	8	9	16	13	9	5	1	2	1
Size 5 |	8	6	12	10	15	13	18	15	8	8
6 |	8	10	10	8	11	16	14	11	13	17
7 |	8	12	11	11	8	6	10	10	7	8
8 |	3	8	8	12	8	12	10	8	7	7
9 |	7	13	5	7	4	3	10	8	6	6
10 |	12	17	9	6	4	4	7	5	6	5

Fig. 4. Rule 3 results for each *Patch Size/Gap Size* combination. Table values (above) represent the number of simulation runs (out of 30) in which the altruistic food sharing allele evolved to fixation in a viable population. Each of the values greater than zero is plotted in the bar chart along the axis labeled *# of Altruistic Fixations*. *None of the runs yielded a viable population at time of allele fixation

condition capable of supporting a viable population. Note that the frequencies at which this version of the altruistic trait evolved to fixation are greater than Rule 1 and Rule 2 frequencies in the transitional range of resource patchiness. In addition, it is apparent that Rule 3 altruistic alleles evolved to fixation more frequently in a much larger range of ecological conditions than did either of the less sophisticated altruistic alleles.

In the final analysis, one must consider how the preliminary results of this baseline model inform our understanding of how altruistic food sharing could have evolved in Plio/Pleistocene hominid populations. First, we have learned that food sharing behaviors need not be overly complicated to evolve within the transitional range of ecological patchiness. Thus, in certain ecological circumstances even rudimentary food sharing behaviors could have evolved to fixation in hominid populations during the Plio/Pleistocene. The premise that early hominids displayed relatively simple food sharing strategies is more parsimonious than the traditional application of modern hunter-gatherer behaviors. According to the preliminary results, the more parsimonious hypothesis could also be more accurate in some environmental scenarios. This conclusion implies that if the earliest food sharing behaviors were indeed simple, a strong temporal correlation should exist between significant forest fragmentation and the spread of this altruistic behavior. Of course, high resolution field research on the timing and spatial structure of Pliocene forest fragmentation in East Africa as well as a methodology by which spatial signatures left by food sharing can be recognized in early archaeological assemblages are required to test for the presence of this correlation. Second, we have learned that cultural sophistication liberates food sharing from a narrow transitional range of ecological patchiness, thereby allowing altruistic alleles to evolve to fixation at higher frequencies in a larger proportion of the environmental state space. This finding implies that had early hominids been capable of practicing a complex version of food sharing, one which involved gossip and/or the punishment of social cheaters, closed habitat fragmentation would have played a greatly diminished role in the biosocial process. Therefore, if the earliest food sharing behaviors were culturally sophisticated, paleoanthropologists should not expect to find a strong temporal correlation between the evolution of food sharing and the fragmentation of Pliocene forests in East Africa.

5 Conclusion

Clearly, the results of SHARE do not (in fact, cannot) *prove* that food sharing spread through early hominid populations due to the fragmentation of closed habitats. However, the preliminary population-genetic findings demonstrate that ecological patchiness can facilitate the evolution of food sharing when both altruistic and selfish behavioral traits are present in artificial societies of foragers. At the very least, these results imply that the new conceptual model of early hominid food sharing, which focuses on woodland fragmentation and trait group benefits, deserves further scrutiny. In order to investigate more detailed research questions, subsequent versions of SHARE will include additional experimental variables such as the presence of meat, mortality due to predation/accident/disease, and polygenic food sharing phenotypes.

The initial success of this model suggests that paleoanthropologists could employ different agent-based models to build inferences for a variety of hominid behaviors as an alternative to applying allegedly representative living human and nonhuman primate behaviors to evolutionarily unique cultural remains. Though referential models can only remind us of the way the world *is*, agent-based models are capable of generating inferences about how the world *could have been*. Behavioral reconstructions of species that lack contemporary analogs will only benefit from this type of approach. The population-genetic results of SHARE elucidate the relationship between two important variables in the hominid food sharing equation: Plio/Pleistocene ecological patchiness and the social savvy of early food sharing behaviors. Paleoecologists and paleoanthropologists can now use the archaeological record to solve this equation for two unknowns.

Acknowledgements

SHARE was built around ECO, and I thank John Pepper for making his code available for this research on food sharing. I am indebted to Steve Lansing, who provided some of the computer hardware used to create and execute SHARE. Steve Kuhn offered insightful advice throughout the modeling endeavor. John Olsen, Jonathan Scholnick, and five anonymous reviewers provided helpful comments on earlier drafts of this paper. An Emil W. Haury Fellowship, awarded by the Department of Anthropology at the University of Arizona, supported this research. Finally, I would like to thank those who organized the 2004 Joint Workshop on Multi-Agent and Multi-Agent-Based Simulation and orchestrated the peer review process.

References

1. Blurton Jones, N. G.: A Selfish Origin for Human Food Sharing: Tolerated Theft. Ethology and Sociobiology (1984) 5:1-3
2. Dart, R. A.: The Osteodontokeratic Culture of *Australopithecus prometheus*. Transvaal Museum, Pretoria (1957)
3. Epstein, J. M., Axtell, R.: Growing Artificial Societies: Social Science from the Bottom-Up. The Brookings Institution Press, Washington, D.C. and MIT Press, Cambridge (1996)
4. Hamilton, W. D.: Innate Social Aptitudes of Man: An Approach from Evolutionary Genetics. In: Fox, R. (ed.): Biosocial Anthropology. Malabry Press, London (1975) 133-155
5. Isaac, G. Ll.: The Food Sharing Behavior of Protohuman Hominids. Scientific American (1978) 238(4):90-108
6. Kohler, T., Gumerman, G. (eds.): Dynamics in Human and Primate Societies. Oxford University Press, Oxford (2000)
7. Lovejoy, C. O.: The Origin of Man. Science (1981) 211:341-350
8. Minar, N., Burkhart, R., Langton, C., Askenazi, M.: The Swarm Simulation System: A Toolkit for Building Multi-agent Systems. SFI Working Papers 96-06-042, Santa Fe (1996)

9. Pepper, J. W., Smuts, B. B.: The Evolution of Cooperation in an Ecological Context: An Agent-Based Model. In: Kohler, T., Gumerman, G. (eds.): Dynamics in Human and Primate Societies. Oxford University Press, Oxford (2000) 45-76

10. Pepper, J. W., Smuts, B. B.: A Mechanism for the Evolution of Altruism among Nonkin: Positive Assortment through Environmental Feedback. American Naturalist (2002) 160(2):205-213

11. Price, G. R.: Selection and Covariance. Nature (1970) 227:520-521

12. Price, G. R.: Extension of Covariance Selection Mathematics. Annals of Human Genetics (1972) 35:485-490

13. Price, G. R.: The Nature of Selection. J. of Theoretical Biology (1995) 175:389-396

14. Sikes, N. E.: Early Hominid Habitat Preferences in East Africa: Paleosol Carbon Isotopic Evidence. J. of Human Evolution (1994) 27:25-45

15. Sober, E., Wilson, D. S.: Unto Others: The Evolution and Psychology of Unselfish Behavior. Harvard University Press, Cambridge (1998)

16. Stern, J. T., Susman, R. L.: The Locomotor Anatomy of *Australopithecus afarensis*. American J. of Physical Anthropology (1983) 60:279-317

17. Susman, R. L., Stern, J. T., Jungers, W. J.: Arboreality and Bipedality in the Hadar Hominids. Folia Primatologica (1984) 43:113-156

18. Tooby, J., DeVore, I.: The Reconstruction of Hominid Behavioral Evolution Through Strategic Modeling. In: Kinzey, W. G. (ed.): The Evolution of Human Behavior: Primate Models. State University of New York Press, Albany (1987) 183-237

19. Vrba, E. S.: Morphology and Environmental Change: How do they Relate in Time? South African J. of Science (1980) 76:61-84

20. Vrba, E. S.: Ecological and Adaptive Changes Associated with Early Hominid Evolution. In: Delsen, E. (ed.): Ancestors: The Hard Evidence. Alan R. Liss, New York (1985) 63-71

21. Wilson, D. S.: Hunting, Sharing, and Multilevel Selection: The Tolerated-Theft Model Revisited. Current Anthropology (1998) 39(1):73-97

22. Wilson, D. S., Pollock, G. B., Dugatkin, L. A.: Can Altruism Evolve in Purely Viscous Populations? Evolutionary Ecology (1992) 6:331-341

23. Wilson, D. S., Sober, E.: Reintroducing Group Selection to the Human Behavioral Sciences. Behavioral and Brain Sciences (1994) 17:585-654

Plant Disease Incursion Management

Lisa Elliston, Ray Hinde, and Alasebu Yainshet

Australian Bureau of Agricultural and Resource Economics, GPO Box 1563,
Canberra ACT 2601, Australia

Abstract. An incursion management model was developed to estimate the regional economy effects of a potential exotic pest or disease incursion in the agricultural sector. By developing an agent based spatial model that integrates the biophysical aspects of the disease incursion with the agricultural production system and the wider regional economy the model can be used to analyze the effectiveness and economic implications of alternative management strategies for a range of different incursion scenarios. A case study application of the model investigates the impact of a potential incursion of Karnal bunt in wheat in a valuable agricultural producing region of Australia.

1 Introduction

Australia has a valued reputation for supplying high quality agricultural products with disease free status to export markets. Disease incursions pose a serious threat to this reputation and could cause significant harm to the agricultural industry and surrounding regional commodities, resulting in considerable losses in trade and incomes.

The exotic incursion management (EIM) model was developed to provide estimates of the direct and indirect costs of plant disease and pest incursions, and to evaluate the strategic and tactical response options available to the government in the event of an exotic incursion in the agricultural sector.

A case study application of the model investigates the impact of a potential incursion of Karnal bunt in wheat. While this disease is known not to occur in Australia currently, its introduction would lead to the immediate and significant loss of valuable export markets, currently valued at more than $3 billion Australian dollars annually. The analysis was undertaken to evaluate a range of alternative post-incursion management strategies. As such, in all scenarios it is assumed that an infestation has, or will, occur.

While a number of studies have quantified the direct costs associated with a Karnal bunt incursion under a range of different scenarios (see [1], [2], and [3]), this work also takes into account the flow-on effects to surrounding regional economies.

2 Exotic Incursion Management Model

The EIM model is an agent based model developed using Cormas, a spatial natural resource and agent based simulation modeling framework based within

P. Davidsson et al. (Eds.): MABS 2004, LNAI 3415, pp. 225–235, 2005.

the VisualWorks programming environment, which allows for the development of applications in the object oriented language, SmallTalk [4].

A number of programming environments can be used to model multi-agent systems. Cormas was selected because it is a platform focused on the building of simulation models where individuals and groups of individuals share, or are located on, a common resource. The established library features, including a strong focus on the visualization of agents and their behavior, facilitate quick model development.

The Cormas development platform contains three modules. In the first module each of the entities within the system are created, and their interactions with other entities are defined. The second module coordinates the overall dynamics, particularly the sequence of events. The third and final module enables the programmer to define the points of view from which the model can be observed.

The model is spatial in nature, with a grid of square cells representing individual paddocks, groups of which are managed as farms by different landholders. Cellular automata techniques are used to drive the spread of the disease across neighboring paddocks, and a range of potential disease transmission vectors are modeled explicitly. At the same time, numerous agents including farmers, contract labor and quarantine officers — each with their own specific patterns of behavior and movement — are also interacting in the spatial environment.

The model consists of a number of distinct components with separate modules capturing the characteristics of the disease; the farming system; the incursion response and management of an outbreak; and a stylized representation of a regional economy to measure the flow-on effects of a disease incursion.

An agent-based approach was used to allow for a more realistic evaluation of alternative management strategies. By explicitly modelling the dynamic responses of farmers to an incursion and any subsequent eradication or control measures in a simulation framework the results provide insight into their likely effectiveness.

2.1   Disease Characteristics

A range of potential disease vectors, identified by plant pathologists as the most likely transmission paths for plant diseases, were explicitly incorporated into the model. Estimates of the spread characteristics were also provided by plant pathologists based on their expert knowledge of the disease (table 1). Each disease vector interacts with the spatial environment, with its own patterns of behavior and movement[1]. For example, contract workers and farmers are able to spread disease across and between farms as they move throughout the region. Subsequently, the disease may spread across neighboring paddocks by wind or the movement of machinery on-farm.

2.2   Farm System

The farm system is modeled as an annual cropping cycle, with each simulation running over 15 years. Each step of the simulation represents the passing of

---

[1] All probabilities are expressed per visit (v), or per season (s).

one week of time. Farmers decide what proportion of their farm to plant to different crops, as well as when to plant them within a planting timeframe that starts in early April and finishes at the end of June. Four agricultural activities are explicitly incorporated in the model: wheat, feed wheat, sorghum and an aggregate representing all other cropping and livestock activities.

After planting, the two major events in the remainder of the year involve spraying for weeds in July and harvesting crops throughout October, November and December. Farmers decide whether they wish to use contractors to spray for weeds and harvest crops, or whether they will do these activities themselves. During these two periods of time the disease can be spread across and between farms in the region. At all other times of the year farmers are randomly moving around their property, potentially spreading the disease.

Based on average yields for the region as reported in the 2001 agricultural census conducted by the Australian Bureau of Statistics (ABS), the total volume of wheat, sorghum and other commodities produced on each farm is calculated [5]. The grain produced is sent to the silo and, based on average farm-gate prices, gross receipts are calculated.

The farmers have not been modeled explicitly as profit maximising agents. There are few agricultural land use alternatives available to farmers in the region analysed so they face a prescribed set of planting options based on their quarantine status.

Table 1. Parameters representing the spread characteristics of the disease

Transmission vector	Probability
Probability of contractor with infected machinery infesting a wheat paddock while spraying for weeds (v)	0.00001
Probability of a contractor with infected machinery infesting a wheat paddock while harvesting (v)	0.5–0.75
Probability of a contractor's machinery becoming infected if they spray for weeds in an infested paddock (v)	0.0001
Probability of a contractor's machinery becoming infected if they harvest an infested paddock (v)	1.0
Probability of a farmer with infected machinery infesting an uninfested wheat paddock elsewhere on their property during harvest time (s)	0.75
Probability of a farmer with infected machinery infesting an uninfested wheat paddock elsewhere on their property at any other time of the year (v)	0.0001–0.75
Probability of a farmer's machinery becoming infected if they are in an infested paddock during harvest time (s)	1.0
Probability of a farmer's machinery becoming infected if they are in an infested paddock at any other time of the year (s)	0.001–0.0001
Probability of disease spreading from one paddock to a neighboring one (due to wind) at harvest time (s)	0.9
Probability of disease spreading from one paddock to a neighboring one (due to wind) at any other time of the year (s)	0–0.25

2.3   Incursion Response and Management

There are two ways in which a disease incursion can be identified: on farm by farmers, or at the silo after harvesting. When the disease is identified a quarantine response is triggered to investigate the extent of the incursion and attempt to contain it so that it cannot spread further. The farm from which the infected grain came is immediately quarantined and a tactical response officer is dispatched to the property. Depending on the specific quarantine rules, a collection of neighboring properties are placed in a buffer quarantine zone. Tactical response personnel visit each neighboring property and search for signs of the disease. If the disease is found on any neighbouring property it is upgraded to full quarantine status. A buffer quarantine zone is established around this farm and all properties within this region are then searched. Where signs of infestation are not found on neighboring properties, those properties remain in the buffer quarantine region and the search of other properties stops.

At the same time, any contractors that have visited infested farms that are now fully quarantined are identified to trace back the source of the incursion and limit its spread. In the first instance, contractors identified in this process are asked to provide a list of all the farms they have visited during the year. Tactical response personnel are then dispatched to each of these farms in order to identify the extent of the incursion. Where an infestation is identified on a property, that property is fully quarantined and the search through all neighboring properties begins. Any contractors contacted in this trace back process who were carrying the disease on their machinery are disinfected before the next season begins.

In the event of an identified disease incursion, farms can be classified as: identified as having an infestation and fully quarantined; identified as not having an infestation but in a buffer quarantine zone because neighboring properties have an infestation; or not quarantined, either clear of infestation or not yet identified.

Depending on the characteristics of the particular disease incursion, a range of control measures can be put in place. These can include the application of chemical treatments to crops, soil or seed; the destruction of crops; and restrictions on the agricultural activities that can be undertaken for a specified period of time. The effectiveness of these control measures can be subject to some uncertainty. However, in the scenarios presented in this paper it is assumed that the control measures are always fully effective. After control measures have been applied all restrictions are lifted from the previously infested farm.

2.4   The Regional Economy

A twelve sector input-output (I–O) model based on data collected by the ABS represents the regional economy [6]. I–O tables contain the supply and demand of goods and services in an economy over a particular period, along with the interdependencies between the industries and associated primary factors of production. Changes in the value of agricultural production as a result of a disease incursion or any subsequent management can therefore be traced through the rest of the regional economy.

The I–O analysis provides estimates of both the direct and indirect impacts of a change in agricultural production resulting from a disease incursion. The direct — or initial — impact captures the changes in wheat and other grain production and any associated changes in employment and income in the directly affected industries, as well as any changes in imports required by these industries. Subsequent changes in all other industries and the directly affected industries form indirect or flow-on impacts.

I–O analysis can overestimate the results of a change in the economy because it does not allow for price induced flexibility between primary factors of production or labour and capital between different commodities. However, the analysis was considered a reasonable approximation for this regional case study where the potential price impacts are likely to be small. For a more detailed description of the model see [7].

3   Karnal Bunt Case Study

Karnal bunt of wheat is caused by the smut fungus *Tilletia indica* Mitra, and despite causing only minor yield losses, infected grains emit a fishy odor and are unfit for human consumption [8] and [9]. It has never been identified in Australian wheat [10]. Karnal bunt could be introduced into Australia through the importing of infected machinery and farm inputs. While the likelihood of this occurring is low, the costs associated with the immediate loss of valuable grain export markets in the event of an infestation would be significant. As a result, it is a disease of particular concern to the Australian wheat industry.

Karnal bunt teliospores have proven resistant to adverse environmental conditions, remaining viable for up to five years in contaminated soil. The primary means of containing the disease is to ban the planting of wheat on affected farms for at least five years.

The model was calibrated to a case study region in eastern Australia that relies heavily on the agricultural sector and produces high quality wheat that attracts a significant price premium on world market. In baseline simulations with no disease the model generated production statistics comparable with data collected as part of the 2001 agricultural census [5].

3.1   Incursion Scenarios

Two different incursion scenarios, with different levels of farmer detection and reporting were analyzed to investigate the importance of early detection on the likelihood of eradicating the disease and the overall economic cost of a Karnal bunt outbreak. A series of 100 simulations were conducted for each scenario and the results presented reflect the average results of those simulations. Sensitivity analysis has not been included due to space constraints.

The first scenario involves a limited and slowly expanding incursion with Karnal bunt introduced into the case study region by contractor equipment. The incursion begins with just two contractors and spreads across the region by the movement of farmers and contractors, as well as the wind.

The second scenario represents a diffuse starting point with potentially rapid expansion, with a load of fertilizer contaminated with Karnal bunt sold throughout the region at the beginning of the simulation.

The measures put in place to contain the Karnal bunt incursion include the destruction of Karnal bunt infested wheat in the first year that it is identified and the establishment of planting restrictions that last for five years. Farmers identified as having the disease are unable to grow grain crops for five years. Neighboring farms that form the buffer zone are unable to grow grain crops in the first year after an incursion is identified. In the remaining years of the five year quarantine period, wheat grown on these farms can only be used for feed purposes within the region.

Further, when the disease is first identified the price for wheat production in the region, even wheat that is free of the disease, is assumed to receive a lower price. The price for wheat only returns to the higher export price after the disease is deemed to have been eradicated from the region on the basis that no new incursions are identified for at least one year.

3.2   Contractor Based Incursion

Two contractor based incursion scenarios were investigated. In the first scenario, the likelihood of infested grain being identified at the silo was assumed to be 50 per cent, and farmers did not report signs of the disease on their property. In the second scenario, the likelihood of detection at the silo remained at 50 per cent and all farmers reported signs of the disease on their property.

When farmers do not report signs of the disease on their property the area infested increases rapidly to more than 14000 hectares by the fourth year of the fifteen year simulation (figure 1). Almost 80 per cent of all infested land is identified and quarantined, the spread of the incursion is curtailed and the area infested reduces to negligible levels by the end of the planning horizon.

In contrast, when farmers report signs of the disease on their property the extent of the incursion is reduced, with a maximum of 5900 hectares infested by the fourth year of the simulation. More than 95 per cent of all infested land is quarantined in the fourth year and the area infested is reduced significantly. Despite a slightly higher level of infestation in this scenario compared with the scenario where no farmers report in the latter years of the simulation, more than 95 per cent of the infested land is quarantined and the likelihood of the disease being eradicated within another five years is high.

The combined effect of the low level of infestation in the region and the ability of quarantine measures to contain an outbreak caused by contractors means that the adverse economic effects of this hypothetical Karnal bunt incursion are relatively minor. Over the fifteen year planning horizon considered, an incursion, even when no farmers report signs of the disease on their property, results in a net loss of production valued at around $58 million [2] in net present value terms,

---

[2] All economic impacts are reported in Australian dollars (2003 prices) and exclude the administrative costs associated with undertaking surveillance and eradication.

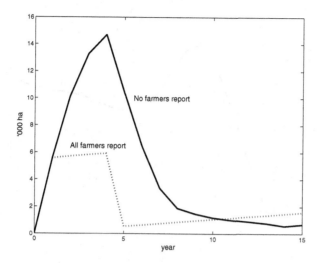

Fig. 1. Area infested, contractor scenario

Table 2. Regional economy effect of alternative incursion scenarios ($AUD)

	Initial (direct)	Flow-on (indirect)	Total
Contractor incursion, no farms report	−58.0	−22.3	−80.4
Contractor incursion, all farms report	−55.6	−21.5	−77.1
Fertilizer incursion, no farms report	−430.2	−165.3	−595.5
Fertilizer incursion, all farms report	−368.9	−141.5	−510.3

which represents around 2 per cent of the value of grain production in the region (table 2).

Over a fifteen year planning horizon, the indirect effect of the hypothetical incursion on all industries is estimated to be around $22 million. The total industry and consumption effects, reflecting the indirect effects along with the initial (direct) effects, capture the overall impact of this particular Karnal bunt incursion. It is estimated that over the fifteen year planning horizon, the decline across the case study region is around $80 million.

When farmers report signs of the disease on their property the incursion is contained in a shorter period of time and the overall economic effects of the outbreak are reduced. Over the fifteen year planning horizon, the loss in value of production is estimated at under $56 million. When the direct and indirect effects of changes in production are aggregated across the region, the decline in economic performance is $77 million.

The $3.3 million difference in the economic performance of the region under these two contractor based incursion scenarios provides an indication of the

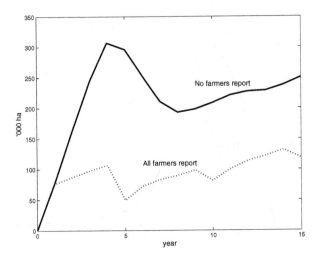

Fig. 2. Area infested, fertilizer scenario

value associated with improving the likelihood of detection by farmers on their property. This in turn can provide a benchmark against which expenditure aimed at improving farmer awareness of the disease, and therefore the likelihood of detection, can be assessed.

3.3   Fertilizer Based Incursion

Two fertilizer based incursions, with the same likelihood of detection at the silo and on-farm, were also investigated. When farmers do not report signs of the disease on their property, the area infested increases to more than 300000 hectares by the fourth year of the simulation (figure 2). Unlike the contractor scenario, only around two-thirds of all infested land is identified and quarantined at this point in the simulation. The disease fails to be contained and the area infested continues to increase throughout the remainder of the planning horizon. When farmers report signs of the disease on their property, the extent of the incursion is reduced significantly, but still fails to be eradicated.

The much larger incidence of infestation across the region and the failure of quarantine measures to adequately contain the disease when it is brought into the region via contaminated fertilizer results in the economic impact of this scenario being much larger than the contractor based incursion scenario.

Over the fifteen year planning horizon, a fertilizer based incursion where farmers do not report signs of the disease results in a net loss of agricultural production valued at around $430 million, compared with a reference case of no disease, which represents around 20 per cent of the value of grain production in the region (table 2). When the indirect effects are added to this, the economic impact of the disease incursion is estimated at more than $595 million over the fifteen year planning horizon.

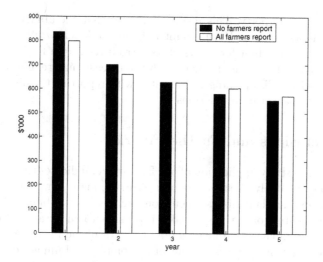

Fig. 3. Receipts per infested farm, contractor scenario

When farmers do report signs of the disease on their property the extent of the incursion is reduced and the economic impact of the incursion is correspondingly reduced. The net loss in agricultural production that results from the fertilizer based incursion falls to around \$369 million when farmers report signs of the disease on their property. This converts to an overall loss in regional income of \$510 million.

The difference in the economic performance of the region under the two fertilizer based incursion scenarios again provides an indication of the value associated with improving the likelihood of detection by farmers on their property.

3.4   Incentives to Self Report

The reporting of the disease by farmers is likely to be a social dilemma as reflected in the *tragedy of the commons*. While the region as a whole is better off when farmers report signs of the disease on their property, it is likely that individual farmers are made worse off by such reporting. See [11] for a detailed review of the agent based simulation of these social dilemmas.

In the contractor based incursion scenario the receipts per infested farm for the scenario where no farmers report signs of the disease on their property are higher than the scenario where all farmers report for the first two years (figure 3). In the third year the receipts per farm are on average equivalent. In subsequent years when the quarantine measures take effect and the disease is contained, the receipts per infested farm are higher under the scenario where farmers report signs of the disease on their property. Similar results were observed for the fertilizer based incursion.

These results indicate that while the region as a whole is better off when farmers report signs of the disease on their property, at least in the early years of the simulation farmers appear to have little incentive to report signs of the disease on their property. This results indicates that the ability to successfully contain a disease such as Karnal bunt is likely to depend critically on it being identified within the first few years of its introduction to a region.

To ensure self reporting, farmers with the disease may require financial payments to offset the disadvantages associated with reporting the disease and being placed under quarantine restrictions.

4   Conclusions and Further Work

The initial case study application of the EIM model indicates that it can be used for assessing the likely effectiveness of alternative management strategies in the event of an exotic plant disease incursion such as Karnal bunt. The integrated bio-economic nature of the model makes it possible to undertake a comprehensive analysis of the impact of an incursion and any resulting incursion management. This includes not only an assessment of the biophysical impact of an incursion, but the resulting on-farm and regional economy effects as well.

The results from the case study analysis demonstrate the ability of improved farmer detection and reporting to reduce the overall economic costs associated with exotic disease incursions. They also highlight the need to provide appropriate incentives to farmers in order to obtain the estimated benefits of eradication or containment to the regional economy.

This preliminary assessment of the incentives that farmers face to engage in strategic behavior highlights the ability of an agent based modeling approach to enhance the accuracy with which the effect of proposed containment strategies can be assessed. Further development on the representation of the economic agents within the model is being undertaken to explicitly capture their incentives to cooperatively reduce the effect of an exotic incursion.

The modeling framework presented is generic in nature and is capable of being adapted to analyze a wide range of incursion scenarios, as well as the incursion of other plant diseases and pests. It is anticipated that additional development of the framework will occur so that the model can be extended to analyze the spread of both weeds and insect pests in other case study regions.

References

1. CIE (Centre for International Economics): A Tale of Two Models: A more detailed look at the Karnal bunt case study, (2002), Report prepared for Plant Health Australia, Canberra.
2. Murray, G.: Pest Risk Analysis on Karnal Bunt of Wheat. Risk Analysis Report, NSW Agriculture, Sydney (1998).
3. Brennan, J.P. and Warham, E.J.: Economic losses from Karnal bunt of wheat in Mexico, CIMMYT Economics Working Paper 90/02, International Maize and Wheat Improvement Centre, Mexico (1990).
4. CIRAD (Centre de coopération internationale en recherche agronomique pour le développement): CORMAS: Natural Resouces and Agent-Based Simulations, (2003), Montpellier, France, (cormas.cirad.fr/indexeng.htm)

5. ABS (Australian Bureau of Statistics): 2001 Agricultural Census, (2001), Canberra, Australia.
6. ABS (Australian Bureau of Statistics): Australian National Accounts, Input–Output Multipliers 1989-90, cat. no. 5237.0, Canberra, Australia.
7. Elliston, L., Hinde, R., Yainshet, A.: Karnal Bunt: The Regional Economic Effects of a Potential Incursion. ABARE eReport 04.4 Prepared for Plant Health Australia, (2004), Canberra, Australia, (freely available for download from www.abareonlineshop.com).
8. Bonde, M., Peterson, G., Schaad, N., Smilanick, J.: Karnal bunt of wheat. Plant Disease 81 (1997) 1370–7.
9. Nagarajan, S., Aujla, S.S., Nanda, G.S., Sharma, I., Goel, L.B., Kumar, J. and Singh, D.V.: Karnal bunt (Tilletia indica) of wheat – a review. Review of Plant Pathology. 76(12) 1207–14.
10. Stansbury, C.D., McKirdy, S.J., Diggle, A.J. and Riley, I.T.: Modeling the risk of entry, establishment, spread, containment and economic impact of Tilletia indica, the cause of Karnal bunt of wheat, using an Australian context. Phytopathology. 92(3) 321–31.
11. Gotts, N., Polhill, J., Law, A.: Agent-Based Simulation in the Study of Social Dilemmas. Artificial Intelligence Review 19 (2003) 3–92.

A Hybrid Micro-Simulator for Determining the Effects of Governmental Control Policies on Transport Chains

Markus Bergkvist[1], Paul Davidsson[1], Jan A. Persson[2], and Linda Ramstedt[2]

[1] Blekinge Institute of Technology, Department Systems and Software Engineering,
372 25 Ronneby, Sweden

[2] Blekinge Institute of Technology, Department Systems and Software Engineering,
374 24 Karlshamn, Sweden
{Markus.Bergkvist, Paul.Davidsson, Jan.Persson, Linda.Ramstedt}@bth.se

Abstract. A simulation-based tool is described which will be used to investigate how the actors in a transport chain are expected to act when different types of governmental control policies are applied, such as, fuel taxes, road tolls, vehicle taxes and requirements on vehicles. The simulator is composed of two layers, one layer simulating the physical activities taking place in the transport chain, e.g., production, storage, and transports of goods, and another layer simulating the different actors' decision making processes. The decision layer is implemented by a multi-agent system where each agent corresponds to a particular actor and models the way it acts in different situations. The simulator will be used for analyzing the costs and environmental effects, and will in this way provide guidance in decision making regarding control policies. In addition, it will be possible for companies to use the simulator in order to determine cost-effective strategies given different (future) scenarios.

1 Introduction

This paper describes a simulation-based tool with the aim to investigate how the actors in a transport chain are expected to act when different types of governmental control policies, such as, fuel taxes, road fees, vehicle taxes and requirements on vehicles, are applied. The policy making is driven by a desire to attain a sustainable environment (by reducing emissions, noise, accidents, and so on) and to achieve sustainable economical development. From a societal perspective, the simulator will be used to analyze the total costs and environmental effects of a transport chain and in this way provide guidance in decision making regarding control policies. The intention is that such analyses will complement those made using existing macro-models. In addition, it will be possible for businesses to use the simulator in order to determine cost-effective strategies given different (future) scenarios.

In the next section we further motive the need for the type of tool suggested and review some related work. We then describe the problem domain, the structure of the simulator, and a small case study. A discussion and pointers to future work concludes the paper.

P. Davidsson et al. (Eds.): MABS 2004, LNAI 3415, pp. 236–247, 2005.

2 Background

The importance of being able to predict the effects of governmental control policies can be illustrated by the *marginal cost principle*. According to this principle, the external costs of transports, such as, emissions, road wear, congestion, noise, accidents etc., should be internalized. It has been argued that the current fees and taxes for heavy transports do not correspond to the actual external costs caused by these transports [5]. To apply the marginal cost principle it is necessary to change some taxes, fees, or regulations. But in order to know which action(s) to take, it is important to have deep knowledge regarding the effects of these actions, i.e., how the different actions will change the behavior of the actors involved in transport chains. This is important in order for policy makers to take a long term perspective supporting sustainable growth of trade and industry. From the perspective of the actors in a transport chain, they need to develop strategies for acting given different future implementations of control policies.

2.1 Existing Simulation Models for Transport Systems

Traditionally, the effects of control policies have been studied using *macro-level* models, such as SAMGODS (SAMPLAN) [9], ASTRA [1] and SISD [6]. These models are taking a societal perspective and are based on aggregated course-grained data on the national level. A problem with these models is that they do not take the logistical processes into account, e.g., choice of carrier type and inventory strategies, and thus fail to model the level where the decisions regarding the actual transports are taking place. Models that take logistical aspects into consideration are for example SMILE [10], GoodTrip [2] and SLAM [6]. However, these models cannot take specific properties of individual transport chains into account. Due to (increased) cooperation between actors in transport chains (e.g., producers, customers, transport operators), there exists a significant flexibility of how to carry out their operations given different control policies. We believe that more precise predictions regarding the effects of control policies can be achieved using micro-level models, i.e., transport chain level models, that capture also the decision making of the actors in the logistical processes.

2.2 Multi Agent Based Simulation for Policy Making

As Multi Agent Based Simulation, MABS, and other micro simulation techniques, explicitly attempts to model specific behaviors of specific individuals, it may be contrasted to macro simulation techniques that are typically based on mathematical (equation-based) models where the characteristics of a population are averaged together and the model attempts to simulate changes in these averaged characteristics for the whole population. Thus, in macro simulations, the set of individuals is viewed as a structure that can be characterized by a number of variables, whereas in micro simulations the structure is viewed as emergent from the interactions between the individuals. According to Parunak et al. [7] "...agent-based modeling is most appropriate for domains characterized by a high degree of localization and distribution and dominated by discrete decision. Equation-based modeling is most naturally applied to systems that can be modeled centrally, and in which the dynamics are dominated by physical laws rather than information processing." Obviously, transport systems fulfill all the characteristics of domains appropriate for agent-based modeling.

As an example of an application of MABS for policy making, consider Downing et al. [4], who have used it in the context of climate policy and climate change. A prototype agent-based integrated assessment model was proposed for water issues like drought, flood etc. where the social relations that support the effectiveness of exhortation are described. Downing et al. argue that MABS is well-suited for this purpose since agents represent the behavior of different actors, here policy makers and households, and the interaction between the agents can therefore be described and evaluated. Also, since MABS can represent different grains, couplings to macro-models can be done.

3 The Problem Domain

The general area investigated is decision support for public policy makers in the area of transportation and traffic. In particular we study the question: What would the consequences be in a transport chain[1] given a certain policy? We envision a simulation-based decision support system were a policy maker is able to experiment with different types of fees, taxes, requirements on vehicles, etc. and get feedback from the system regarding the predicted effects of these policies.

The consequences of public control policies on transport chains are closely connected to the decisions made within the chain, such as, choosing mode of transportation, carrier size, when to transport, which quantities to transport. These decisions are made by different actors at different levels in the chain and may have implications on the system which are rather hard to anticipate.

In general a transport chain can be organized in a number of different ways with respect to the owner of the products at different locations and to the decision makers organizational belonging, e.g., the transport could be carried out by either the seller, buyer or third party logistics operator. The decision making in transport chains is subject to both short- and long-term planning implying that the time dimension of the decisions needs to be considered when modeling the transport chains. We will assume that the decision makers (actors) are cost minimizers locally with only minor exploration of potential cost savings achievable by cooperation in the transport chain. This appears to be rather typical in transport chains today, e.g. a customer orders a certain quantity to be delivered at a particular time and date. However, we also plan to incorporate market-based cooperation between the simulated actors allowing for a behavior which approaches a system optimal behavior. This represents the ongoing development in transport chains, e.g. negotiations of when and of which quantities to deliver to the customer occurs in order to reduce costs of production and transportation.

---

[1] We avoid the term supply chain, since it implies indirectly that a customer view is taken, i.e. to supply a customer, rather than a system view. Also, our focus is on transports, whereas production and consumption provides the context in which the transports take place.

The input to the simulator is:

- the transport tasks, i.e., a sequence of customer requirements
- the available transport resources and their characteristics, such as costs, capacity, and environmental performance
- the available production resources and their characteristics
- the available infrastructure, e.g., road and rail networks
- the location of producers, customers, storages, etc.

Given this task, the user of the simulator will be able to experiment with different control policies, by varying a number of parameters corresponding to different taxes, fees, regulations etc.

The output will then include (among other things):

- performed transport operations
- the estimated external costs (including environmental costs)
- society revenue (from taxes, fees, etc)
- the internal costs
- customer satisfaction measured in terms of reliability of deliveries and quality of products.

4 The Simulator

We have chosen a hybrid approach, where an agent-based approach has been used to simulate the decision making activities, and a more traditional object-oriented micro-level approach has been used to simulate the physical activities. This is illustrated in Fig 1 and further described in the remaining part of this section.

4.1 Physical Simulator

The physical simulator is based on the description of the production and distribution network suggested by Davidsson and Wernstedt [3]. It simulates the physical level of the production and distribution of commodities, whereas the decisions for what to produce, where to store the commodities, fleet management, etc. are simulated by the decision making simulator.

There are four basic types of entities in the simulator that makes up the production and distribution: nodes, links, transport carriers and commodities. A *node* is a producer, an internal distribution node, or a customer, and has the following attributes:

- production capacity for each commodity,
- production level (dynamic, i.e., the value may change during the simulation),
- storage capacity (volume) for each commodity type,
- inventory level (dynamic),
- load time for each carrier type, and
- unload time for each carrier type.

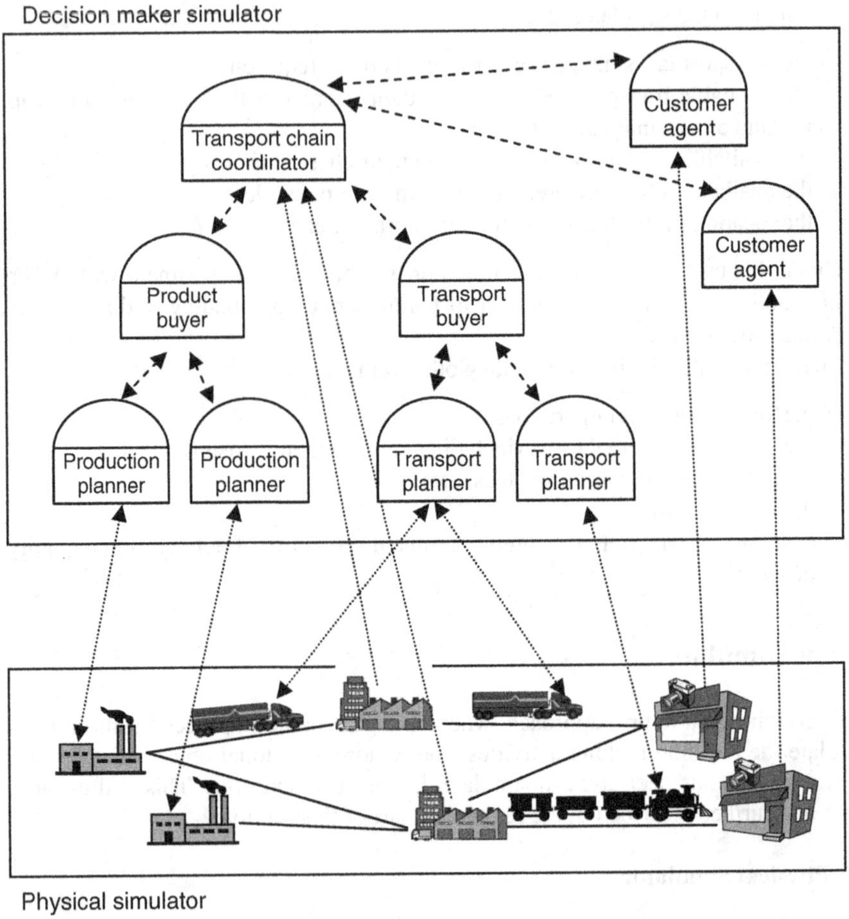

Decision maker simulator

Physical simulator

Fig. 1. The two layers of the simulator.

A *link* connects a pair of nodes in the distribution network and acts as a distribution channel for the transport carriers. A link has the following attributes:

- connected pair of nodes,
- mode of transportation,
- length, and
- average distribution speed.

A *transport carrier* is an entity that performs a transport along a link and has the following attributes:

- carrier type (Each type is associated with a particular mode of transportation.),
- volume capacity for each commodity type,
- location (dynamic),
- load (dynamic),
- maximum speed,

- delay probability distribution,
- transport cost, and
- environmental performance.

A *commodity* is produced at nodes and transported via links by transport carriers and have the following attributes:

- commodity type (based on storage requirements),
- production cost,
- production time,
- mass,
- volume, and
- quality (dynamic, based on age).

The activities in the physical simulator can be controlled during run-time through a number of commands. There are commands available to start a production batch, load and unload commodities from a transport carrier, initiate a transport or consume commodities. Commands that are sent to the simulator are placed locally at the target entity in a first-in-first-out queue.

The available commands, their constraints and expected outcomes are:

- *Manufacture(n, c, s)*. Adds a new command to the command queue of node n to start a new production batch of commodity c of size s. The command is executed if the node has the required production capacity. The time until the batch is completed is determined by the production time. When the batch is completed the new commodities are placed in storage at n.
- *Load(v, c, s, n)*. Adds a new command to the command queue of transport carrier v to load the quantity s of commodity c from the storage of node n. The command is executed if the transport carrier is located at node n. It then requests the commodities from the node which returns the commodities (if available) and the time it takes to load them.
- *Unload(v, c, s, n)*. Adds a new command to the command queue of transport carrier v to unload the quantity s of commodity c to the storage of node n. Works similar to the Load command with the difference that a request to unload is sent to the node.
- *Dispatch(v, e)*. Adds a new command to the command queue of transport carrier v to initiate a transport using link e. The command can only be executed if the carrier is at either of the nodes connected by e, and is not un/loading.
- *Consume(n, c, s)*. Adds a new command to the command queue of the node n to consume quantity s of commodity c from the storage of node n.

In addition, it is possible to read the attributes of all entities.

4.2 Decision Making Simulator

We have identified a number of important roles in a transport chain, which are described in Table 1.

Table 1. The modeled roles of a transport chain

Decision maker	Decisions and actions	Based on	Goal
Customer agent	Makes requests of products with respect to quantities, time of delivery (or time window), and quality level.	Anticipated customer demand and inventory levels at customer.	Mediate customer requirements in the most accurate way that is possible.
Transport chain coordinator	Decides how much should be bought from producers and how much should be taken from storages. Makes requests to product and transport buyers.	Requests from the customer agents, intermediate inventory levels, and transport and production opportunities.	Satisfy the customer requirements at the lowest possible cost.
Product buyer	From which producer should the products be bought? Makes request of production to production planners.	Requests from transport chain coordinator. Bids from producers (production planners), including prices, deadlines, quality of product etc.	Satisfy the product requirements at the lowest price possible (given the constraints).
Production planner	What is the best bid that the producer can provide? Gives production orders to the producer.	Production capacity, storage levels (at the production site).	Minimize production costs.
Transport buyer	From which transport operator should the transport be bought? Makes request of transports to transport planners.	Requests from transport chain coordinator. Bids from transport operators (transport planners), including prices, quality of transport, etc.	Satisfy the transport requirements at the lowest price possible (given the constraints).
Transport planner	What is the best bid that the transport operator can provide? Assigns tasks to transport carriers (fleet management).	Status (availability, position, etc) of the transport carriers controlled by the operator.	Minimize transport costs.

There are many possible mappings between organizations and the different decision making agents. In the extreme case, all decision makers belong to the same organization for a transport chain, e.g., petroleum companies. Another extreme, is where all decision makers belongs to different organizations. Also, intermediate arrangements exist such as one of the real world cases in the project. Below, some mappings between agents and organizations are suggested:

- The customer agent might be a retailer or a producer, with the goal to buy a desired quantity of goods to the lowest price delivered at a desired time. However, this agent can typically accept (to a reduced price) to receive the products earlier than required and hence store the product until needed.
- The transport chain coordinator might, for example, be a planner within a larger company or a third or fourth party logistics operator.
- The product buyer is often connected to the organization which hosts the transport chain coordinator. However, it can be independent, for example, in case the transport chain coordinator is a third party logistics operator.
- The production planner belongs typically to the producing company.
- The transport buyer might belong to different type of organizations, for instance, the transport buyer might belong to the same organization as the customer or as the transport chain operator.
- The transport planner typically belongs to the organization owning and controlling the transport carriers.

As a case study we have selected a transport chain within the food industry consisting of: Karlshamns AB, a producer of fats and oils, the transport operator FoodTankers, and a typical buyer of bulk products from Karlshamns AB, Procordia Foods. A number of the decision making agents, i.e., the transport chain coordinator, product buyer, production planner, and transport buyer, are all associated with Karlshamns AB. The customer and the transport planner agents are associated with Procordia Foods and FoodTankers, respectively.

The suggested hierarchical design of the decision maker simulator allows for the study of different levels of cooperation. It allows for modeling the extreme (but rather common) case, where the agents have pure local objectives (local cost minimizer) with virtually no sharing of information. Further, the design allows for the other extreme case, where the agents are fully cooperative with the objective of minimizing total cost of the system. In order to approach system optimality, however, an optimization mechanism needs to be applied for guiding the decision agents. In Persson and Davidsson [8], an example of such a mechanism is outlined for a similar problem setting.

The decision making simulator primarily models operational decisions. Strategic decisions, such as buying or selling of vehicles, increase or decrease of storage capacity, are not explicitly modeled. However, these decisions may indirectly be accounted for by the user of the simulator, or directly in a more advanced version of the simulator by extending the decision domain.

5 A Simple Case Study

In order to illustrate the usage of the simulator we have chosen to describe a very simple case study. The scenario consists of a producer of fluids (density 1000 kg/m^3) in Karlshamn, situated in southern Sweden, a customer in Fredrikstad situated in southern Norway, and no internal distribution nodes. There are two transport operators available, one using trucks and one using rail. We focus on the transport selection task, assuming that sufficient amount of products and carriers are available to meet the demands of the customers. Details of the two *links* that connects the two nodes and the three different *transport carrier* types used in the case study are given in Table 2 and 3 respectively.

Table 2. The links of the transport chain

	Link A	Link B
Nodes	Karlshamn, Fredrikstad	Karlshamn, Fredrikstad
Mode	Road	Rail
Length (km)	540	600
Average speed (km/h)	72	14

Table 3. The attributes of the transport carrier types. The cost and environmental performance for trucks depends on the load [empty, full]. The values for rail are based on proportions of the size of the average freight train set in Sweden which is 535 tonnes (according to the Swedish Network for Transport and Environment, see www.ntm.a.se)

	Truck	Rail_27	Rail_50
Mode	Road	Rail	Rail
Volume capacity (m^3)	30	27	50
Max speed (km/h)	90	90	90
Probability of delay	0	0	0
Cost (€)	[665, 680]	1005	1764
Env. perf.: CO_2 (g/km)	[754, 891]	143	265

Further assumptions and characteristics of the studied scenario are given below.

- A time horizon 52 weeks is considered.
- Two customer orders are generated per week. The time and the quantity of the order are randomly generated. The quantities are generated to match an ideal size of a truck, which is 30 tonnes with a probability of 0.5 or of a rail freight carriage of either 27 or 50 tonnes, with a probability of 0.25, respectively.
- In the scenario, the effects (costs and environmental performance) of returning the truck or the rail freight carriage to the producer have been ignored. This has been ignored in these initial experiments since the effects are highly dependent on the possibility to take on other loads on the return trip which is not modeled explicitly.
- It is assumed that the train carriage is transported using diesel engines for 30% of the distance and electrical engines for 70% of the distance, since only parts of the railway network are electrified.
- It is assumed that products can be loaded directly into and directly from the different types of vehicles (train and truck) at both the customer and the producer.

We have chosen to study the effect of using kilometer taxation on heavy trucks, since this governmental control policy is under discussion for implementation in Sweden. In Figures 2 and 3, the results have been plotted for different levels of kilometer taxation. We study the effect on the total cost for the customer, tax income, and emitted carbon dioxide. As expected, the cost is increasing for increasing kilometer taxation. The tax income is not zero for zero kilometer taxation, since a fuel tax is associated with the diesel of €0.55 which applies both for trucks and diesel trains.

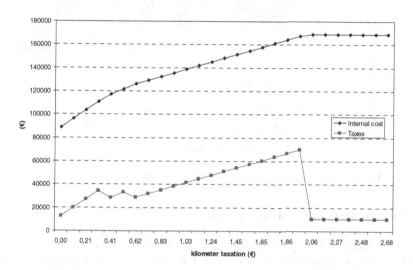

Fig. 2. Internal costs and taxes for different values of kilometer taxation

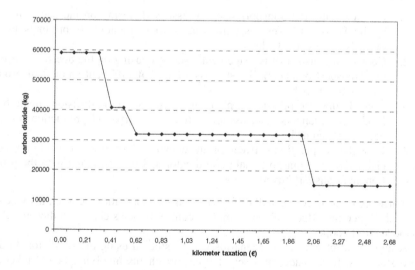

Fig. 3. Carbon dioxide emissions

In Figure 4, the transition from using only trucks and to using only train is illustrated. Using only trucks is competitive up to a kilometer taxation of €0.38; and in order to make only trains competitive a kilometer taxation of €2.0 is required. These breakpoints will naturally shift if the cost for moving vehicles back to the producer is fully considered.

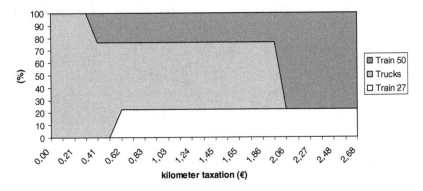

Fig. 4. Percentage of the three carrier types for different kilometer taxations

6   Conclusions and Future Work

We have outlined a hybrid micro-level simulator which is currently being developed. It models the physical activities as well as decision making activities taking place in transport chains. We will use this simulator in order to study the effects of different governmental control policies. This is done in several steps with increasing complexity:

1. Consider only the transport selection issues (focusing on the decisions made by the Transport buyer), assuming that sufficient amount of products and carriers always are available.
2. Considering also fleet management issues (focusing on the decisions made by the Transport planners), but still assuming that sufficient amount of products always are available.
3. Considering also production planning (focusing on the decisions made by the Production planners), but assuming that transports and production are independently planned.
4. Consider cooperation between producers and transporters (focusing on the decisions of the Transport chain coordinator and its interaction with the production and transport buyers).

Once step four is completed, different levels of cooperation between actors can be studied. Hence the effects of governmental control policies can be studied in relation to the level of cooperation.

A number of characteristic scenarios will be studied using the simulator. In-depth case studies will be made on a set of transport chains involving FoodTankers and Karlshamns AB (see below). Validation of the simulation model will be carried out partly through these case studies, and partly though close cooperation with SIKA (the Swedish Institute for Transport and Communications Analysis). SIKA has much experience in the area and access to vast amount of data concerning the Swedish transport and traffic systems. As they also are a potential user of the simulator, they are participating in the requirements analysis.

Acknowledgements

This work is carried out within the project "Effects of Governmental Control Policies in Transportation Chains: A Micro-level Study" (see www.ipd.bth.se/stem), which is mainly financed by VINNOVA, the Swedish Agency for Innovation Systems. Further, the Swedish Knowledge Foundation is in part supporting this research, via the project "Integrated Production and Transportation Planning within Food Industry" (see www.ipd.bth.se/fatplan) Two companies are primarily involved in the case study, FoodTankers which provides services in the form of tanker transport, and Karlshamns AB which is one of the world's leading producers of vegetable oils and fats. In addition, the project is supported by an expert committee including experts from SIKA and the Swedish Environmental Protection Agency (via the MiSt research programme).

References

1. ASTRA – Assessment of Transport Strategies, Final Report, *Deliverable to EU DGTREN*, ASTRA Consortium, Karlsruhe, ASTRA, 2000. (Available at http://www.iww.uni-karlsruhe.de/ASTRA/ASTRA_Final_Report.pdf)
2. Boerkamps, J., van Binsbergen, A., GoodTrip – A New Approach for Modelling and Evaluation of Urban Goods Distribution, Urban Transport Conference, 2nd KFB Research Conference, Lund, Sweden, 7-8 June 1999.
3. Davidsson, P., and Wernstedt, F.: A framework for evaluation of multi-agent system approaches to logistics network management. *Multi-Agent Systems: An Application Science*, Kluwer, 2004.
4. Downing, T.E., Moss, S., Pahl-Wostl, C.: Understanding Climate Policy Using Participatory Agent-Based Social Simulation, In Moss, S. and Davidsson, P. (eds.) *Multi-Agent-Based Simulation*, Springer, 2001.
5. Hesselborn P.-O., Swahn, H., Ekonomiska styrmedel i transportpolitiken – Förslag till utveckling av den svenska modellen, SIKA Dokument 1998:1, Presented at the University of Aalborg at the conference "Trafikdage i Aalborg"
6. ME&P – WSP, DfT Integrated Transport and Economic Appraisal, Review of Freight Modelling, 2002 (http://www.dft.gov.uk/stellent/groups/dft_transstrat/ documents/page/dft_transstrat_508058.pdf)
7. Parunak H.V.D., Savit R., Riolo R.L.: Agent-Based Modeling vs. Equation-Based Modeling: A Case Study and Users' Guide. In Sichman, J.S., Conte, R., Gilbert, N. (eds.), *Multi-Agent Systems and Agent-Based Simulation*, pp. 10-26, Springer, 1998.
8. Persson J. A. and Davidsson, P.: Integrated optimization and multi-agent technology for combined production and transportation planning, *HICSS-38, Hawaii International Conference on Systems Science*, 2005.
9. Swahn, H.: The Swedish model systems for goods transport – SAMGODS. A brief introductory overview. *SAMPLAN Report 2001:1*, SIKA, 2001.
10. Tavasszy, L.A., van de Vlist, M., Ruijgrok, M., van de Rest, J.: Scenario-wise analysis of transport and logistics systems with a SMILE, *8th World Conference on Transport Research*, Antwerp, Belgium, 1998.

Simulation and Analysis of Shared Extended Mind

Tibor Bosse[1], Catholijn M. Jonker[1], Martijn C. Schut[1], and Jan Treur[1,2]

[1] Vrije Universiteit Amsterdam, Department of Artificial Intelligence,
De Boelelaan 1081a, NL-1081 HV Amsterdam, The Netherlands
{tbosse, jonker, schut, treur}@cs.vu.nl
http://www.cs.vu.nl/~{tbosse, jonker, schut, treur}
[2] Utrecht University, Department of Philosophy,
Heidelberglaan 8, 3584 CS Utrecht

Abstract Some types of animals exploit patterns created in the environment as external mental states, thus obtaining an extension of their mind. In the case of social animals the creation and exploitation of such patterns can be shared, which supports a form of shared extended mind or collective intelligence. This paper explores this shared extended mind principle for social animals in more detail. Its main goal is to analyse and formalise the dynamic properties of the processes involved, both at the local level (the basic mechanisms) and the global level (the emerging properties of the whole), and their relationships. A case study in social ant behaviour in which shared extended mind plays an important role is used as illustration. For this case simulations are described based on specifications of local properties, and global properties are specified and verified.

1 Introduction

In [4], [5], [6], [8], [14], [15] it is described how behaviour is often not only supported by an internal mind in the sense of internal mental structures and cognitive processes, but also by processes based on patterns created in the external environment that serve as external mental structures. Examples of this pattern of behaviour are the use of 'to do lists' and 'lists of desiderata'. Having written these down externally (e.g., on paper, in your diary, in your organiser or computer) makes it unnecessary to have an internal memory about all the items. Thus internal mental processing can be kept less complex. The only thing to remember is where these lists are available. Other examples of the use of extended mind are doing mathematics or arithmetic, where external (symbolic, graphical, material) representations are used; e.g., [3].

Clark and Chalmers [6] point at the similarity between cognitive processes in the head and some processes involving the external world. This similarity can be used as an indication that these processes can be considered extended cognitive processes or extended mind: 'If, as we confront some task, a part of the world functions as a process which, *were it done in the head*, we would have no hesitation in recognizing as part of the cognitive process, then that part of the world *is* (so we claim) part of the cognitive process. Cognitive processes ain't (all) in the head!' [6], Section 2. (...)

P. Davidsson et al. (Eds.): MABS 2004, LNAI 3415, pp. 248–264, 2005.

'Of course, one could always try to explain my action in terms of internal processes and a long series of "inputs" and "actions", but this explanation would be needlessly complex. If an isomorphic process were going on in the head, we would feel no urge to characterize it in this cumbersome way.' [6], Section 3. As the patterns in the external world have to be created and sensed, interaction with the external world will be more intensive, compared to the case where internal mental states are created and exploited.

Especially in the case of social animals external mental states created by one individual can be exploited by another individual, or, more generally, the creation, maintenance, and exploitation of external mental states are activities in which a number of individuals can participate (for example, presenting slides on a paper with multiple authors to an audience). Further examples can be found everywhere, varying from roads and traffic signs to books or other media, and to many other kinds of cultural achievements. In this multi-agent case the extended mind principle serves as a way to build a form of social or collective intelligence that goes beyond (and may even not require) social intelligence based on direct one-to-one communication. In such cases the external mental states cross and, in a sense, break up the borders between (the minds of) the individuals and become shared mental states.

An interesting and currently often studied example of collective intelligence is the intelligence shown by ant colonies [1], [7]. Indeed, in this case the external world is exploited as an extended mind by using pheromones. While they walk, ants drop pheromones on the ground. The same or other ants sense these pheromones and follow the route in the direction of the strongest concentration. Because pheromones evaporate, such routes may vary over time.

The goal of this paper is to analyse this shared mind principle in more detail, and to provide a formalisation of its dynamics. These are illustrated by a case study of social behaviour based on shared extended mind (a simple ant colony). The analysis of this case study comprises multi-agent simulation based on identified local dynamic properties, identification of dynamic properties for the overall process, and verification of these dynamic properties.

More specifically, Section 2 is a brief introduction of the basic concepts used in the modelling approach and formalisation. It introduces two modelling languages, one (the *leads to* language) used for simulation, and one (the Temporal Trace Language TTL) for more complex properties that can be used in analysis. For the former language a software environment for simulation has been developed, for the latter language a software environment has been developed that enables automatic checking of specified properties against given traces. In Section 3 a simulation model is presented for the ant case study. This simulation model is specified using local properties: temporal rules that express in a local manner the basic mechanisms of the case. These rules are specified and formalised in the *leads to* language introduced in Section 2, and are therefore directly executable in the software environment that has been developed. Some of the simulation outcomes are included in Section 3. Whereas Section 3 has a local perspective on the basic mechanisms, Section 4 takes the global perspective of emergent properties of the multi-agent process as a whole. A number of relevant global dynamic properties are identified and formalised in the language

TTL. It is discussed how these global dynamic properties have been checked against simulation traces. Moreover, some of the logical relationships between them are discussed. Section 5 is a discussion of the results.

2 State Properties and Dynamic Properties

Dynamics will be described in the next section as evolution of states over time. The notion of state as used here is characterised on the basis of an ontology defining a set of physical and/or mental (state) properties that do or do not hold at a certain point in time. For example, the internal state property 'the agent A has pain', or the external world state property 'the environmental temperature is 7° C', may be expressed in terms of different ontologies. To formalise state property descriptions, an ontology is specified as a finite set of sorts, constants within these sorts, and relations and functions over these sorts. The example properties mentioned above then can be defined by nullary predicates (or proposition symbols) such as pain, or by using n-ary predicates (with n≥1) like has_temperature(environment, 7). For a given ontology Ont, the propositional language signature consisting of all *state ground atoms* (or *atomic state properties*) based on Ont is denoted by APROP(Ont). The *state properties* based on a certain ontology Ont are formalised by the propositions that can be made (using conjunction, negation, disjunction, implication) from the ground atoms. A *state* S is an indication of which atomic state properties are true and which are false, i.e., a mapping S: APROP(Ont) → {true, false}.

To describe the internal and external dynamics of the agent, explicit reference is made to time. Dynamic properties can be formulated that relate a state at one point in time to a state at another point in time. A simple example is the following informally stated dynamic property for belief creation based on observation:

'if the agent observes at t1 that it is raining, then the agent will believe that it is raining'.

To express such dynamic properties, and other, more sophisticated ones, the Temporal Trace Language TTL is used; cf. [10]. In this language, explicit references can be made to time points and traces. Here a *trace* or *trajectory* over an ontology Ont is a time-indexed sequence of states over Ont. The sorted predicate logic temporal trace language TTL is built on atoms referring to, e.g., traces, time and state properties. For example, 'in trace γ at time t property p holds' is formalised by state(γ, t) |= p. Here |= is a predicate symbol in the language, usually used in infix notation, which is comparable to the Holds-predicate in situation calculus. Dynamic properties are expressed by temporal statements built using the usual logical connectives and quantification (for example, over traces, time and state properties). For example, consider the following dynamic property:

'in any trace γ, if at any point in time t1 the agent A observes that it is raining, then there exists a time point t2 after t1 such that at t2 in the trace the agent A believes that it is raining'.

In formalised TTL form it looks as follows:

∀t1   [state(γ, t1) |= observes(A, itsraining)   ⇒
       ∃t2 ≥ t1   state(γ, t2) |= belief(A, itsraining)   ]

Language abstractions by introducing new (definable) predicates for complex expressions are possible and supported.

In order to specify simulation models, a simpler temporal language has been developed, based on TTL. This language (the *leads to* language) enables one to model direct temporal dependencies between two state properties in successive states. This executable format is defined as follows. Let α and β be state properties of the form 'conjunction of atoms or negations of atoms', and e, f, g, h non-negative real numbers. In the *leads to* language $\alpha \twoheadrightarrow_{e, f, g, h} \beta$, means:

If state property α holds for a certain time interval with duration g,
then after some delay (between e and f) state property β will hold for a certain
time interval of length h.

For a precise definition of the *leads to* format in terms of the language TTL, see [11]. A specification of dynamic properties in *leads to* format has as advantages that it is executable and that it can often easily be depicted graphically.

3   A Simulation Model of Shared Extended Mind

Dynamic properties can be specified at different aggregation levels, varying from (local) dynamic properties for the basic mechanisms and (global) properties of a process as a whole. This section introduces the local dynamic properties for the basic mechanisms; they are used to specify a simulation model. The world in which the ants live is described by a labeled graph as depicted in Figure 1.

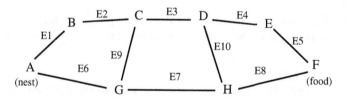

Fig. 1. An ants world

Locations are indicated by A, B,..., and edges by E1, E2,... The ants move from location to location via edges; while passing an edge, pheromones are dropped. The objective of the ants is to find food and bring this back to their nest. In this example there is only one nest (at location A) and one food source (at location F).

The example concerns multiple agents (the ants), each of which has input (to observe) and output (for moving and dropping pheromones) states, and a physical body which is at certain positions over time, but no internal mental state properties (they are assumed to act purely by stimulus-response behaviour). An overview of the formalisation of the state properties of this single agent conceptualisation is shown in Table 1. In these local properties, a is a variable that stands for ant, l for location, e for edge, and i for pheromone level.

Table 1. Formalisation of state properties

	body positions in world:
pheromone level at edge e is i	pheromones_at(e, i)
ant a is at location l coming from e	is_at_location_from(a, l, e)
ant a is at edge e to l2 coming from location l1	is_at_edge_from_to(a, e, l1, l2)
ant a is carrying food	is_carrying_food(a)
	world state properties:
edge e connects location l1 and l2	connected_to_via(l1, l2, e)
location l is the nest location	nest_location(l)
location l is the food location	food_location(l)
location l has i neighbours	neighbours(l, i)
edge e is most attractive for ant a coming from location l	attractive_direction_at(a, l, e)
	input state properties:
ant a observes that it is at location l coming from edge e	observes(a, is_at_location_from(l, e))
ant a observes that it is at edge e to l2 coming from location l1	observes(a, is_at_edge_from_to(e, l1, l2))
ant a observes that edge e has pheromone level i	observes(a, pheromones_at(e, i))
	output state properties:
ant a initiates action to go to edge e to l2 coming from location l1	to_be_performed(a, go_to_edge_from_to(e, l1, l2))
ant a initiates action to go to location l coming from edge e	to_be_performed(a, go_to_location_from(l, e))
ant a initiates action to drop pheromones at edge e coming from location l	to_be_performed(a, drop_pheromones_at_edge_from(e, l))
ant a initiates action to pick up food	to_be_performed(a, pick_up_food)
ant a initiates action to drop food	to_be_performed(a, drop_food)

To model the example a number of local dynamic properties are used, of which a small subset is provided in this section. The complete set of local properties used is given in Appendix A.

LP5 (Selection of Edge)

This property models (part of) the edge selection mechanism of the ants. It expresses that, when an ant observes that it is at location l, and there are two edges connected to that location, then the ant goes to the edge with the highest amount of pheromones. Formalisation:

observes(a, is_at_location_from(l, e0)) and neighbours(l, 3) and connected_to_via(l, l1, e1) and observes(a, pheromones_at(e1, i1)) and connected_to_via(l, l2, e2) and observes(a, pheromones_at(e2, i2)) and e0 ≠ e1 and e0 ≠ e2 and e1 ≠ e2 and i1 > i2 •→» to_be_performed(a, go_to_edge_from_to(e1, l1))

LP9 (Dropping of Pheromones)
This property expresses that, if an ant observes that it is at an edge e from a location l to a location l1, then it will drop pheromones at this edge e. Formalisation:
observes(a, is_at_edge_from_to(e, l, l1)) •→→ to_be_performed(a, drop_
pheromones_at_edge_from(e, l))

Fig. 2. Simulation trace with three ants, part I

LP13 (Increment of Pheromones)
This property models (part of) the increment of the number of pheromones at an edge as a result of ants dropping pheromones. It expresses that, if an ant drops pheromones at edge e, and no other ants drop pheromones at this edge, then the new number of pheromones at e becomes i*decay+incr. Here, i is the old number of pheromones, decay is the decay factor, and incr is the amount of pheromones dropped. Formalisation:

to_be_performed(a1, drop_pheromones_at_edge_from(e, I1)) and ∀I2 not to_be_performed(a2, drop_pheromones_at_edge_from(e, I2)) and ∀I3 not to_be_performed(a3, drop_pheromones_at_edge_from(e, I3)) and a1 ≠ a2 and a1 ≠ a3 and a2 ≠ a3 and pheromones_at(e, i) •↠ pheromones_at(e, i*decay+incr)

LP14 (Collecting of Food)
This property expresses that, if an ant observes that it is at location F (the food source), then it will pick up some food. Formalisation:

observes(a, is_at_location_from(l, e)) and food_location(l) •↠ to_be_ performed(a, pick_up_food)

LP18 (Decay of Pheromones)
This property expresses that, if the old amount of pheromones at an edge is i, and there is no ant dropping any pheromones at this edge, then the new amount of phero-mones at e will be i*decay. Formalisation:

pheromones_at(e, i) and ∀a,l not to_be_performed(a, drop_pheromones_ at_edge_from(e, l)) •↠ pheromones_at(e, i*decay)

A special software environment has been created to enable the simulation of ex-ecutable models. Based on an input consisting of dynamic properties in *leads to* for-mat, the software environment generates simulation traces. Example of such traces can be seen in Figure 2, 3 and 4. Time is on the horizontal axis, the state properties are on the vertical axis. A dark box on top of the line indicates that the property is true during that time period, and a lighter box below the line indicates that the property is false. This trace is based on all local properties identified. Because of space limita-tions, in the example depicted in Figure 2 and 3, only three ants are involved. The trace in Figure 4 shows an example with two ants. However, similar experiments have been performed with a population of 50 ants. Since the abstract way of modelling used for the simulation is not computationally expensive, also these simulations re-quired no more than 30 seconds each.

Figure 2 and 3 are both parts from the same trace. Figure 2 shows the observations and locations of the ants; Figure 3 shows the performed actions of the ants. As can be seen in Figure 2 and 3, there are two ants (ant1 and ant2) that start their search for food immediately, whereas ant3 comes into play a bit later, at time point 3. When ant1 and ant2 start their search, none of the locations contain any pheromones yet, so basically they have a random choice where to go. In the current example, ant1 selects a rather long route to the food source (via locations A-B-C-D-E-F), whilst

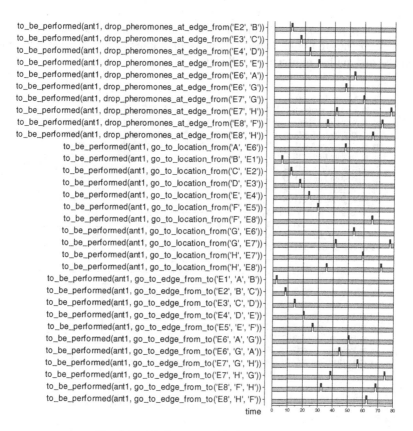

Fig. 3. Simulation trace with three ants, part II

ant2 chooses a shorter route (A-G-H-F). Note that, in the current model, a fixed route preference (via the attractiveness predicate) has been assigned to each ant for the case there are no pheromones yet. After that, at time point 3, ant3 starts its search for food. At that moment, there are trails of pheromones leading to both locations B and G, but these trails contain exactly the same number of pheromones. Thus, ant3 also has a choice among location B and G, and chooses in this case to go to B. Meanwhile, at time point 18, ant2 has arrived at the food source (location F). Since it is the first to discover this location, the only present trail leading back to the nest, is its own trail. Thus ant2 will return home via its own trail. Next, when ant1 discovers the food source (at time point 31), it will notice that there is a trail leading back that is stronger than its own trail (since ant2 has already walked there twice: back and forth, not too long ago). As a result, it will follow this trail and will keep following ant2 forever. Something similar holds for ant3. The first time that it reaches the food source, ant3 will still follow its own trail, but some time later (from time point 63) it will also follow the other two ants. To conclude, eventually the shortest of both routes is shown to remain, whilst the other route evaporates. Other simulations, in particular for small ant populations, show that it is important that the decay parameter of the pheromones is not too high. Otherwise, the trail leading to the nest has evaporated before the first ant has returned, and all ants get lost.

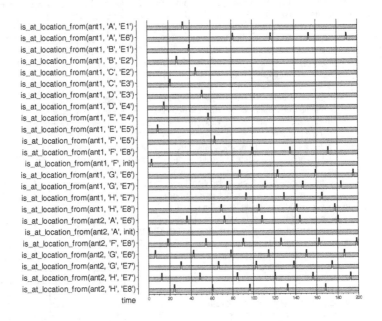

Fig. 4. Simulation trace with two ants, starting at different locations

In Figure 4, there is one ant (ant1) that starts its search departing from the food location and one ant (ant2) that starts slightly later departing from the nest location. The first ant (ant1) takes the long way home (via locations F-E-D-C-B-A), while the second ant (ant2) immediately takes the short route (via locations A-G-H-F) to the food. Figure 4 shows that after some time, both ants follow the short route. Thus also for this example, we may conclude that eventually the shortest of both routes is shown to remain, whilst the other route evaporates.

4 Global Properties and Verification

In the previous section dynamic properties at the lowest aggregation level (the local dynamic properties) were addressed, and simulation based on these properties was discussed. The current section addresses dynamic properties of a global nature and their verification. Within these properties, γ is a variable that stands for an arbitrary trace. First a language abstraction is given:

food_delivered_by(γ, t, a) $\equiv$ $\exists$l, e [state(γ,t) |= is_at_location_from(a, l, e)) &
state(γ,t) |= nest_location(l) & state(γ,t) |= to_be_performed(a, drop_food)]

GP1 Food Delivery Succesfulness
There is at least one ant that brings food back to the nest.

$\exists t \exists a$: food_delivered_by(γ, t, a).

GP2 Multiple Delivery
Food is delivered by more than one ant

$\exists t1$, t2 $\exists a1$, a2 [a1 $\neq$ a2 & food_delivered_by(γ, t1, a1) & food_delivered_by(γ, t2, a2)]

Other language abstractions are:

attractive_route_to(γ, a, x) $\equiv$
$\exists l \, \exists e \, \forall t$ [state(γ, t) $\models$ attractive_direction_at(a, l, e) & state(γ, t) $\models$ connected_ o_via(l, x, e)]

i.e., the attractive route of ant a passes through location x.

reaches_end_attractive_route(γ, t, a) $\equiv$
$\exists l$, e [state(γ, t) $\models$ is_at_location_from(a, l, e) &
attractive_route_to(γ, a, l) & $\forall e'$ state(γ, t) $\models$/$=$ attractive_direction_at(a, l, e')]

GP3 Reaching End of Attractive Route
Ants reach the end of their attractive route.

$\forall a \, \exists t$ reaches_end_attractive_route(γ, t, a)

GP4 Returning To Nest
Ants get back to the nest from the end of their attractive routes.

$\forall a \, \forall t1 \, \exists e$, t2 > t1 $\exists l$ [reaches_end_attractive_route(γ, t1, a) $\Rightarrow$
state(γ, t2) $\models$ is_at_location_from(a, l, e) & state(γ, t2) $\models$ nest_location(l)]

GP5 From Food To Nest
Ants get back to the nest from locations of food.

$\forall a$, e $\forall t1 \, \exists t2 > t1 \, \exists l$, l', e'
[state(γ, t1) $\models$ is_at_location_from(a, l, e) & state(γ, t1) $\models$ food_location(l)] $\Rightarrow$
state(γ, t2) $\models$ is_at_location_from(a, l', e') & state(γ, t2) $\models$ nest_location(l')

These and a number of other properties have been formalised and using a checking software environment have been (automatically) *verified in simulation traces*. This is a first manner for verification. A second way of verification is to establish *logical relationships* between properties (by mathematical proof). This also has been performed in a number of cases. For example, under a number of assumptions the following relationships hold:

GP4 $\Rightarrow$ GP5
GP3 & GP4 $\Rightarrow$ GP2

The assumptions include:

- attractive routes are not branching and are not crossing each other or themselves.
- at least two ants exist for which the attractive routes end at a food location and are short enough compared to the evaporation rate of pheromones to return.
- GP5 is only valid in the infinite future, since food sources are not depleted. In practice, the simulations stop, invalidating GP5 for the ants that are still on their way to the nest.

Furthermore, an additional premise of Temporal Completion, see [9], is needed. For example, any of the following trivial (non-intended) world situations would disturb the ants: an ant comes to a location that contains a pheromone that is there without any reason (no ant dropped it), or on its way back an ant comes to a location without a pheromone (the pheromone immediately disappeared). It is clear that the above properties can only be proven under the assumption that nothing unexpected will happen. To put it differently, proofs can be given under the assumption that the set of local properties determines the whole range of events. This assumption has been added as a premise to establish the logical relationships between the properties.

5 Discussion

Clark and Chalmers [6], Section 5, provide four criteria for an extended mind: (1) the external information is a constant in the agent's life - when the information is relevant, he will rarely take action without consulting it; (2) the external information is directly available; (3) the agent endorses retrieved external information; (4) the external information has been endorsed at some point in the past, and is there as a consequence of this endorsement. How do these criteria apply to the ants case? First, indeed an ant always senses the pheromone before choosing a direction. Second, at each location the pheromone is immediately accessible for sensing. Third, the decision for the direction is indeed always based on the pheromone. Finally, the external information is endorsed in the past: the pheromone was dropped at the direction from whence one or more ants traveled.

The extended mind perspective introduces an additional, cognitive ontology to describe properties of the physical world, which essentially is an antireductionist step, providing a more abstract and better manageable, higher level conceptualisation. In [12] a number of arguments can be found of why such antireductionist steps can be useful in explanation and theory development. Indeed, following the extended mind perspective a high-level conceptualisation was obtained. This high-level conceptualisation could be formalised and analysed in a logical manner. The formalisation enables simulation and automated checking of dynamic properties of traces or sets of traces, and allows one to logically relate dynamic properties of different aggregation levels to each other. All this would have been more difficult in the case of an algorithmic or physically-oriented modelling perspective, involving, for example, differential equations and gradients of concentrations.

As an extension to the current paper, in [2] the notion of *representational content* is analysed for mental processes based on the shared extended mind principle. The analysis of notions of representational content of *internal* mental state properties is well-known in the literature on Cognitive Science and Philosophy of Mind. In this literature a relevant internal mental state property m is taken and a representation relation is identified that indicates in which way m relates to properties in the external world or the agent's interaction with the external world; cf. [13, pp. 184-210]. For the case of extended mind an extension of the analysis of notions of representational content to *external* state properties is needed. Moreover, for the case of external mental state properties that are *shared*, a notion of *collective* representational content is needed (in contrast to a notion of representational content for a single agent).

With respect to the simulation, work is currently in progress to replace the behaviour prescribed by attractiveness of a route by random route selection. In addition, experiments with food sources at different distances from the nest will be undertaken to determine the relation between evaporation rate and ants finding their way home. Therefore, these food sources will be made depletive. Also, the effect of using different types of pheromones will be studied. Moreover, an advanced visualisation environment is currently developed to make the simulation traces more readable.

Finally, work is in progress to addresses the question in how far a process involving multiple agents that shows some form of collective intelligence can be interpreted as a single agent. Like in the current paper, this question will be answered by formal analysis. It will be explored for example processes how they can be conceptualised and formalised in two different manners: from a single agent or from a multi-agent perspective. Moreover, an ontological mapping will be formally defined between the two formalisations, in order to show how collective behaviour can be interpreted as single agent behaviour.

Acknowledgements

The authors are grateful to Lourens van der Meij for his contribution to the development of the software environment.

References

1. Bonabeau, J. Dorigo, M. and Theraulaz, G. (1999). *Swarm Intelligence: From Natural to Artificial Systems*. Oxford University Press, New York, 1999.
2. Bosse, T., Jonker, C.M., Schut, M.C., and Treur, J. (2004). Modelling Shared Extended Mind and Collective Representational Content. *Proceedings of the 24th International Conference on Innovative Techniques and Applications of Artificial Intelligence*. Lecture Notes in AI, Springer Verlag. To appear.
3. Bosse, T., Jonker, C.M., and Treur, J. (2002). Simulation and analysis of controlled multi-representational reasoning processes. *Proc. of the Fifth International Conference on Cognitive Modelling, ICCM'03*. Universitats-Verlag Bamberg, 2003, pp. 27-32.

4. Clark, A. (1997). *Being There: Putting Brain, Body and World Together Again.* MIT Press, 1997.
5. Clark, A. (2001). Reasons, Robots and the Extended Mind. In: *Mind & Language,* vol. 16, 2001, pp. 121-145.
6. Clark, A., and Chalmers, D. (1998). The Extended Mind. In: *Analysis,* vol. 58, 1998, pp. 7-19.
7. Deneubourg, J.L., Aron S, Goss S., Pasteels J. M. and Duerinck G. (1986). Random Behavior, Amplification Processes and Number of Participants: How They Contribute to the Foraging Properties of Ants. In: *Evolution, Games and Learning: Models for Adaptation in Machines and Nature,* North Holland, Amsterdam, 1986, pp. 176-186.
8. Dennett, D.C. (1996). *Kinds of Mind: Towards an Understanding of Consciousness,* New York: Basic Books.
9. Engelfriet, J. Jonker, C.M., and Treur, J. (2002). Compositional verification of Multi-Agent Systems in Temporal Multi-Epistemic Logic. *Journal of Logic, Language and Information,* vol. 11, 2002, pp. 195-225.
10. Jonker, C.M., and Treur, J., Compositional Verification of Multi-Agent Systems: a Formal Analysis of Pro-activeness and Reactiveness. *International Journal of Cooperative Information Systems,* vol. 11, 2002, pp. 51-92.
11. Jonker, C.M., Treur, J., and Wijngaards, W.C.A., A Temporal Modelling Environment for Internally Grounded Beliefs, Desires and Intentions. *Cognitive Systems Research Journal,* vol. 4(3), 2003, pp. 191-210.
12. Jonker, C.M., Treur, J., and Wijngaards, W.C.A., Reductionist and Antireductionist Perspectives on Dynamics. *Philosophical Psychology Journal,* vol. 15, 2002, pp. 381-409.
13. Kim, J. (1996). *Philosophy of Mind.* Westview Press
14. Kirsh, D. & Maglio, P. (1994). On distinguishing epistemic from pragmatic action. *Cognitive Science,* vol. 18, 1994, pp. 513-49.
15. Menary, R. (ed.) (2004). *The Extended Mind,* Papers presented at the Conference *The Extended Mind - The Very Idea: Philosophical Perspectives on Situated and Embodied Cognition,* University of Hertfordshire, 2001. John Benjamins, 2004, to appear.

Appendix A - The Simulation Model

LP1 (Initialisation of Pheromones)
This property expresses that at the start of the simulation, at all locations there are 0 pheromones. Formalisation:

start •↠ pheromones_at(E1, 0.0) and pheromones_at(E2, 0.0) and pheromones_at(E3, 0.0) and pheromones_at(E4, 0.0) and pheromones_at(E5, 0.0) and pheromones_at(E6, 0.0) and pheromones_at(E7, 0.0) and pheromones_at(E8, 0.0) and pheromones_at(E9, 0.0) and pheromones_at(E10, 0.0)

LP2 (Initialisation of Ants)
This property expresses that at the start of the simulation, all ants are at location A. Formalisation:

start •↠ is_at_location_from(ant1, A, init) and is_at_location_from(ant2, A, init) and is_at_location_from(ant3, A, init)

LP3 (Initialisation of World)

These two properties model the ants world. The first property expresses which locations are connected to each other, and via which edges they are connected. The second property expresses for each location how many neighbours it has. Formalisation:

start •–» connected_to_via(A, B, I1) and ... and connected_to_via(D, H, I10)
start •–» neighbours(A, 2) and ... and neighbours(H, 3)

LP4 (Initialisation of Attractive Directions)

This property expresses for each ant and each location, which edge is most attractive for the ant at if it arrives at that location. This criterion can be used in case an ant arrives at a location where there are two edges with an equal amount of pheromones. Formalisation:

start •–» attractive_direction_at(ant1, A, E1) and ... and attractive_direction_at(ant3, E, E5)

LP5 (Selection of Edge)

These properties model the edge selection mechanism of the ants. For example, the first property expresses that, when an ant observes that it is at location A, and both edges connected to location A have the same number of pheromones, then the ant goes to its attractive direction. Formalisation:

observes(a, is_at_location_from(A, e0)) and attractive_direction_at(a, A, e1) and connected_to_via(A, I1, e1) and observes(a, pheromones_at(e1, i1)) and connected_to_via(A, I2, e2) and observes(a, pheromones_at(e2, i2)) and e1 \= e2 and i1 = i2 •–» to_be_performed(a, go_to_edge_from_to(e1, A, I1))

observes(a, is_at_location_from(A, e0)) and connected_to_via(A, I1, e1) and observes(a, pheromones_at(e1, i1)) and connected_to_via(A, I2, e2) and observes(a, pheromones_at(e2, i2)) and i1 > i2 •–» to_be_performed(a, go_to_edge_from_to(e1, A, I1))

observes(a, is_at_location_from(F, e0)) and connected_to_via(F, I1, e1) and observes(a, pheromones_at(e1, i1)) and connected_to_via(F, I2, e2) and observes(a, pheromones_at(e2, i2)) and i1 > i2 •–» to_be_performed(a, go_to_edge_from_to(e1, F, I1))

observes(a, is_at_location_from(I, e0)) and neighbours(I, 2) and connected_to_via(I, I1, e1) and e0 ≠ e1 and I ≠ A and I ≠ F •–» to_be_performed(a, go_to_edge_from_to(e1, I, I1))

observes(a, is_at_location_from(I, e0)) and attractive_direction_at(a, I, e1) and neighbours(I, 3) and connected_to_via(I, I1, e1) and observes(a, pheromones_at(e1, 0.0)) and connected_to_via(I, I2, e2) and observes(a, pheromones_at(e2, 0.0)) and e0 ≠ e1 and e0 ≠ e2 and e1 ≠ e2 •–» to_be_performed(a, go_to_edge_from_to(e1, I, I1))

observes(a, is_at_location_from(I, e0)) and neighbours(I, 3) and connected_to_via(I, I1, e1) and observes(a, pheromones_at(e1, i1)) and connected_to_via(l, l2, e2) and observes(a, pheromones_at(e2, i2)) and e0 ≠ e1 and e0 ≠ e2 and e1 ≠ e2 and i1 > i2 •↠ to_be_performed(a, go_to_edge_from_to(e1, l1))

LP6 (Arrival at Edge)
This property expresses that, if an ant goes to an edge e from a location l to a location l1, then later the ant will be at this edge e. Formalisation:

to_be_performed(a, go_to_edge_from_to(e, l, l1)) •↠ is_at_edge_from_to(a, e, l, l1)

LP7 (Observation of Edge)
This property expresses that, if an ant is at a certain edge e, going from a location l to a location l1, then it will observe this. Formalisation:

is_at_edge_from_to(a, e, l, l1) •↠ observes(a, is_at_edge_from_to(e, l, l1))

LP8 (Movement to Location)
This property expresses that, if an ant observes that it is at an edge e from a location l to a location l1, then it will go to location l1. Formalisation:

observes(a, is_at_edge_from_to(e, l, l1)) •↠ to_be_performed(a, go_to_location_from(l1, e))

LP9 (Dropping of Pheromones)
This property expresses that, if an ant observes that it is at an edge e from a location l to a location l1, then it will drop pheromones at this edge e. Formalisation:

observes(a, is_at_edge_from_to(e, l, l1)) •↠ to_be_performed(a, drop_pheromones_at_edge_from(e, l))

LP10 (Arrival at Location)
This property expresses that, if an ant goes to a location l from an edge e, then later it will be at this location l. Formalisation:

to_be_performed(a, go_to_location_from(l, e)) •↠ is_at_location_from(a, l, e)

LP11 (Observation of Location)
This property expresses that, if an ant is at a certain location l, then it will observe this. Formalisation:

is_at_location_from(a, l, e) •↠ observes(a, is_at_location_from(l, e))

LP12 (Observation of Pheromones)
This property expresses that, if an ant is at a certain location l, then it will observe the number of pheromones present at all edges that are connected to location l. Formalisation:

is_at_location_from(a, l, e0) and connected_to_via(l, l1, e1) and phero-mones_at(e1, i) •↠ observes(a, pheromones_at(e1, i))

LP13 (Increment of Pheromones)
These properties model the increment of the number of pheromones at an edge as a result of ants dropping pheromones. For example, the first property expresses that, if

an ant drops pheromones at edge e, and no other ants drop pheromones at this edge, then the new number of pheromones at e becomes i*decay+incr. Here, i is the old number of pheromones, decay is the decay factor, and incr is the amount of pheromones dropped. Formalisation:

to_be_performed(a1, drop_pheromones_at_edge_from(e, l1)) and ∀l2 not to_be_performed(a2, drop_pheromones_at_edge_from(e, l2)) and ∀l3 not to_be_performed(a3, drop_pheromones_at_edge_from(e, l3)) and a1 ≠ a2 and a1 ≠ a3 and a2 ≠ a3 and pheromones_at(e, i) •→» pheromones_at(e, i*decay+incr)

to_be_performed(a1, drop_pheromones_at_edge_from(e, l1)) and to_be_performed(a2, drop_pheromones_at_edge_from(e, l2)) and ∀l3 not to_be_performed(a3, drop_pheromones_at_edge_from(e, l3)) and a1 ≠ a2 and a1 ≠ a3 and a2 ≠ a3 and pheromones_at(e, i) •→» pheromones_at(e, i*decay+incr+incr)

to_be_performed(a1, drop_pheromones_at_edge_from(e, l1)) and to_be_performed(a2, drop_pheromones_at_edge_from(e, l2)) and to_be_performed(a3, drop_pheromones_at_edge_from(e, l3)) and a1 ≠ a2 and a1 ≠ a3 and a2 ≠ a3 and pheromones_at(e, i) •→» pheromones_at(e, i*decay + incr+incr+incr)

LP14 (Collecting of Food)
This property expresses that, if an ant observes that it is at location F (the food source), then it will pick up some food. Formalisation:

observes(a, is_at_location_from(l, e)) and food_location(l) •→» to_be_performed(a, pick_up_food)

LP15 (Carrying of Food)
This property expresses that, if an ant picks up food, then as a result it will be carrying food. Formalisation:

to_be_performed(a, pick_up_food) •→» is_carrying_food(a)

LP16 (Dropping of Food)
This property expresses that, if an ant is carrying food, and observes that it is at location A (the nest), then the ant will drop the food. Formalisation:

observes(a, is_at_location_from(l, e)) and nest_location(l) and is_carrying_food(a) •→» to_be_performed(a, drop_food)

LP17 (Persistence of Food)
This property expresses that, as long as an ant that is carrying food does not drop the food, it will keep on carrying it. Formalisation:

is_carrying_food(a) and not to_be_performed(a, drop_food) •→» is_carrying_food(a)

LP18 (Decay of Pheromones)
This property expresses that, if the old amount of pheromones at an edge is i, and there is no ant dropping any pheromones at this edge, then the new amount of pheromones at e will be i*decay. Formalisation:

pheromones_at(e, i) and ∀a,l not to_be_performed(a, drop_pheromones_
at_edge_from(e, l)) ●→» pheromones_at(e, i*decay)

Author Index

Lecture Notes in Artificial Intelligence (LNAI)